Ethics in Information Technology

Cognitive Approaches in Cloud and Edge Computing

Series Editors:
Niranjanamurthy M and Ramiz Aliguliyev

Privacy and Security Challenges in Cloud Computing
A Holistic Approach
T. Ananth Kumar, T. S. Arun Samuel, R. Dinesh Jackson Samuel and
M. Niranjanamurthy

Ethics in Information Technology
A Practical Guide
G. K. Awari and Sarvesh V. Warjurkar

For more information about this series, please visit: www.routledge.com/Cognitive-Approaches-in-Cloud-and-Edge-Computing/book-series/CACAEC

Ethics in Information Technology
A Practical Guide

G. K. Awari and Sarvesh V. Warjurkar

CRC Press
Taylor & Francis Group
Boca Raton London New York

CRC Press is an imprint of the
Taylor & Francis Group, an **informa** business

First edition published 2022
by CRC Press
6000 Broken Sound Parkway NW, Suite 300, Boca Raton, FL 33487-2742

and by CRC Press
4 Park Square, Milton Park, Abingdon, Oxon, OX14 4RN

© 2022 G. K. Awari and Sarvesh V. Warjurkar

CRC Press is an imprint of Taylor & Francis Group, LLC

Library of Congress Cataloging-in-Publication Data
A catalog record has been requested for this book

ISBN: 978-1-032-24975-9 (hbk)
ISBN: 978-1-032-16379-6 (pbk)
ISBN: 978-1-003-28098-9 (ebk)

DOI: 10.1201/9781003280989

Typeset in Times New Roman
by codeMantra

The book is dedicated to **Late Smt Shashikala Awari** *and*
Late Mr. Sheshraoji Warjurkar *for their*
heavenly blessings and encouragement.

Contents

Foreword

It has been my pleasure to know Dr. G. K. Awari and Mr. Sarvesh Warjurkar as my mentor and a friend. Dr. G. K. Awari was the Principal of the college where I was working as an Assistant Professor in the year 2014–2016 and Mr. Sarvesh Warjurkar was my colleague during this time.

Information Technology's influence is now felt all over the world, and it's critical that this impact be researched from various angles. This technology, particularly the Internet, has brought to a never-before-seen clash of ideas. International trade and business, as well as international interactions in general, have needed some cultural knowledge, although few individuals were involved prior to the Internet.

This book *Ethics in Information Technology: A Practical Guide* is increasingly focusing on "people-ware", organizational, behavioral, and social changes that come with the spread of information technology in healthcare settings. The book's authors are prominent specialists who have provided quantum of information technology breakthroughs. The sections in the book increasingly address the importance of information in driving the development of the IT sector throughout the world, rather than only hardware and software. The history of hacking, differenst sorts of hackers, different types of hacking assaults, key hacking tools and software, and concealing IP addresses are all discussed in this book. It also speaks about mobile hacking, hacking an email address, penetration testing, and spoofing attacks.

Cyber Criminals, Mystery, Danger, Money Technology, Cyber Terrorist, while all of these things together sound like the makings of a best-selling fiction novel, the cyber security industry and all of the threats and dangers that exist within it is all too real. That's one reason why cyber security and ethics books make for some pretty interesting reading both in terms of academics and entertainment. Good cyber security and ethics books share insights gained from real-world situations and examples that we can learn from as professionals. It's the great ones that teach us what to look out for so that we're prepared to prevent ourselves from falling prey to cybercriminals.

Apart from that, the authors concentrate on Intellectual Property, which is a relatively new topic. You own homes, jewelry, stocks, cash, and other assets. These are real-world items. You take pleasure in them because you own them. These items that you own cannot typically be taken away and enjoyed by others unless and until you offer them to someone else. This book provides a quick overview of the many aspects of intellectual property and how to protect them from possible exploitation.

This book also includes case studies on topics such as privacy, online trust, anonymity, values-sensitive design, machine ethics, professional behavior, and software developers' moral duty that have fuelled debate about information computing ethics during the last three decades. It teaches you how to comprehend and analyze practical, moral, and legal issues that affect your job and personal life.

Both authors have excellent credentials to develop a work of this kind. Each has published widely on multidisciplinary approaches in the ethics of Information Technology as well as in other areas, and has a deep knowledge of both Eastern and

Western traditions. The chapters reflect this. *Ethics in Information Technology: A Practical Guide* will be an important book that will extend the boundaries of the field of Information Technology ethics and, hopefully, will stimulate much more discussion of this kind. The world of the early twenty-first century needs it.

Dr. Ankush Maind
Ph.D. (VNIT Nagpur)
Lead Solution Consultant
PERSISTENT SYSTEMS,
Nagpur (M.S.), India

Preface

The book contains comprehensive treatment of Information technology ethics, which is the study of the ethical issues arising out of the use and development of electronic technologies. It envelopes all illustrations and applications of ethics in an IT organization, including typical examples from an examination point of view in various universities across the globe. The book addresses the challenges emerging due to the extensive use of online education systems and provides all levels of readers to get an insight into ethics in information technology. The text uniquely addresses practical guide to these challenges without being ignoring any relevant topic. It is mainly aimed at PG/UG/Diploma in Engineering Courses offered in several Universities. This book will serve as a major resource for students of Computer Science and Engineering, Information Technology, Computer Technology, and Master of Computer Science.

At the end of each chapter, multiple choice questions and review questions have been added to make the book a comprehensive unit in all respects. This book is also useful to prepare the readers for the competitive examinations such as GATE, IES, UPSC, and other public sector undertaking.

The main features of this book are as follows:

i. The subject of ethics in information technology is itself is in innovation phase and it is presented presuming that the reader's goal is to achieve an in-depth knowledge of the subject, which can be applied in real practice.

ii. The book is written in the luculent language, which can be understood by the students of even diploma in engineering courses (junior level) as well as stakeholders of online education in the schools.

iii. The book will help readers perceive the legal, ethical, and societal implications of information technology. This offers updated and momentous coverage of issues such as file sharing, infringement of intellectual property, security risks, Internet crime, identity theft, employee surveillance, privacy, compliance, social networking, and ethics of IT corporations.

iv. This book offers a magnificent foundation in ethical decision-making for modern and imminent business managers and IT professionals.

v. The book consists of block diagrams and self-explanatory figures, photographs, and best practices developed through experiences, and it is inspiring for budding entrepreneurs and all users.

Acknowledgements

The achievement of a mission is never a solo effect; it is the product of the important involvement of a variety of individuals in a direct or indirect way that has enabled us to make it a success. We would like to extend our appreciation and recognize that the guiding lights have imbued us with the right element and have helped us to accomplish this mission. The Book on Ethics in Information Technology: A Practical Guide is the culmination of the authors' classroom and practical experiences.

We are grateful to Hon. Dr. Abhay Wagh, Director of Technical Education (DTE), Mumbai, Dr. Vinod Mohitkar, Director of MSBTE, Mumbai, Dr. Manoj Daigavane, Joint Director of Technical Education, RO, Nagpur and Principal, Government Polytechnic, Nagpur, and Dr. M. R. Chitlange, Secretary, MSBTE, Mumbai for their constant inspiration and encouragement.

We are indebted to Dr. Mohan Gaikwad-Patil, Chairman, Gaikwad-Patil Group of Institutions, Nagpur (MS), Prof. Sandeep Gaikwad, Treasurer Gaikwad-Patil Group of Institutions, Nagpur (MS), Dr. Geeta Padole-Gaikwad, Principal, Dr. P. S. Kadu, Director (IIIL), Prof. Pragati Patil, Vice Principal, Dr. N. V. Chaudhari, Dean (R&D), Prof. Anup Gade, Dean Academics at Tulsiramji Gaikwad-Patil of college and engineering, Technology, Nagpur for extending the computing and laboratories facilities of institute to complete this project. We are gratified by the Dr. Ashok Bagul, Senior Police Inspector, Cyber crime branch, Nagpur, Mr. Deepak Dhote, Operations Manager, IT NetworkZ, Nagpur, for their support in developing the case studies. We are also thankful to Ms. Akansha Bhandarkar and Ms. Prajakta Dahate from TGPCET for assisting us to develop the figures with Corel Draw software. We have gained greatly in preparing the manuscript of this book by referring to several articles, journals and online sources and the open source material. We express our gratitude to all such authors, publications and publishers, many of whom have been included in the bibliography. If someone is left out unintentionally, we will seek their forgiveness.

The authors are very grateful to Dr. Jaji Varghese, Aryabhat Polytechnic, New Delhi, Dr. S. Velumani, Velarar College of Engineering, Erode (TN), Dr. Abhijeet Digalwar, Professor, BITS Pilani, Dr. D. K. Parbat, Government Polytechnic, Bramhapuri, Dr. D. N. Kongre, Prof. Tirpude, Prof. Kumbhar from Government Polytechnic, Nagpur, Dr. Manoj Chandak, Professor, Ramdevbaba college of Engineering, Nagpur, Dr. Sonekar, Professor, JD College of Engineering, Nagpur, Prof. Supratim Saha, from Subharti Institute of Technology and Engineering, Swami Vivekanand Subharti University, Subhartipuram, Meerut, Dr. Amey Bhangale, University of California, Riverside, Dr. V. M. Thakre from SG Amravati University, Amravati, for their constant support and consistent assistance in creating this practical guide of Ethics in IT.

We sincerely thank our mentors, Dr. D. G. Wakde, Dr. L. B. Bhuyar, Dr. G. V. Gotmare, Dr. S. W. Rajurkar and Dr. Sonali Ridhorkar, who have helped us and have been a source of inspiration. We are obliged to cyber cell of Maharashtra Police for their backing. We thank CRC Press, Taylor and Francis Group, especially

Gauravjeet Singh Reen Senior Commissioning Editor-Engineering, Ergonomics and Human Factors, Occupational Health and Safety CRC Press, Taylor and Francis Group, Mr. Lakshay Gaba, Mr. Niranjanamurthy M and Mr. Ramiz Aliguliyev, Series Editor Part of the Cognitive Approaches in Cloud and Edge Computing, CRC Press who has kept our morale high, helped us in preparing and maintaining our schedules, for facilitating the work, regular updates and stood behind us patiently during this entire work.

Indeed, we are thankful to our family members for their timely support in all the efforts of this book, without which this book would not have seen the light of day. The efforts of Mrs. Jaya Awari and Master Vedant Awari, Mr. Keshavrav Awari, Prof. Vijay Warjurkar, Mrs. Sucheta Warjurkar, Mr. Jayesh Warjurkar & Mrs. Ankita Warjurkar, baby Anvi are appreciated for encouraging us to develop the manuscript of the book. Last but most important, we bow our heads to the majesty of the Almighty God and our parents for making our experience one of the most technologically satisfying moments of our lives.

We hope that the book will serve the intent of its readers and that we will continue to receive their help and suggestions. Suggestions to enhance the quality and style of the book are always welcome and accepted and integrated in future editions of the book.

Authors

G. K. Awari, has done his Ph.D. in Mech Engg and has more than 30 years of teaching experience at UG/PG/Diploma and Research levels and 10 years as an approved Principal of A+ Grade NAAC Accredited Engineering Institute. He has been nominated by AICTE as "Margdarshak for NBA Accreditation" of mentee institutes.

He has more than 30 Scopus Indexed Journal papers and 4 Patents, 10 Copyrights, to his name. Recently, he has been awarded with Patent titled as *"Hybrid Tool Concept for Boring, Reaming & Chamfering in a Single Tool"* and the Commercial Product is being developed for Mahindra and Mahindra, Nagpur. He has executed more than 8 funded projects of AICTE as a Principal Investigator and organized 10 ten AICTE-ISTE-SRM University funded National Workshops and 10 International Conferences as a convener.

He has contributed to the development of Academics/Curriculum as Board of Study (BOS) Member at Goa University, SG Amravati University, YCCE, Nagpur, and RTM Nagpur University. He is presently BOS member in GH Raisoni University, Saikheda, MP, Delhi; Skill and Entrepreneurship University, New Delhi; and Chairman BOS, Automobile Engineering at Government Polytechnic (GP), Nagpur.

He has authored a total of 10 books in Engineering and Technology domain, and the following 3 books are authored for CRC Press, Taylor and Francis Group:

1. *Quantitative Techniques in Business, Management and Finance: A Case Study Approach*
2. *Automotive Systems: Principles and Practice*
3. *Additive Manufacturing and 3D Printing Technology: Principles and Practice*

Dr. Awari believes in the Co-learning Process and Participative Management and currently is Head of Automobile Engineering Department at Government Polytechnic, Nagpur.

Sarvesh V. Warjurkar is pursuing Ph.D. in Computer Science & Engineering and has more than 10 years of teaching and industrial experience. His areas of interest are Ethics related to Information Technology, Cyber Security along with Algorithms and Machine Learning. Presently he is working as a Head of Department of Information Technology, Tulsiramji Gaikwad Patil College of Engineering & Technology, Nagpur (M.S.), India.

1 An Overview of Ethics

LEARNING OBJECTIVES

- To understand how different people see their own interests in business circumstances.
- To enable readers to form their own, well-considered opinions on corporate ethical problems.
- To encourage the readers to become more deliberate, disciplined thought while attempting to resolve ethical concerns in business.
- To introduce the notion of corporate social responsibility (CSR) and discuss how it relates to ethical business practices.
- To equip readers to contribute to the ethical considerations of organizations with which they may be associated in a positive way.
- To look at the ethical responsibilities and ideals that exist in the interaction between employers and workers.

1.1 DEFINITION OF ETHICS

Ethics is a set of values that define what is and is not acceptable behavior in a certain society. Ethical behavior is described by a set of widely agreed standards, many of which are almost universal. Lies and stealing, for example, are immoral. Opinions of what constitutes ethical action differ dramatically. For example, software piracy, which is the act of making unauthorized copies of software, is considered ethical in some countries but not ethical in others. With a piracy rate of 94%, Vietnam continues to be the region's most piracy-prone nation. China was the country with the second highest piracy rate, at 92%. The United States (27.9%), Japan (21%), and New Zealand (22%) consistently rank among the countries with the lowest rates of piracy.

Children master complex tasks when they are awake, such as walking, communicating, swimming, and writing the alphabet, which they repeat for the rest of their lives. People often build patterns that help them distinguish between what culture finds good and poor. A virtue is a habit that encourages people to do what is right, and a sin is a habit that encourages people to do what is wrong. Virtues such as fairness and honesty are some of the examples of virtues, while envy and rage are examples of vices. In other words, ethics is a subset of philosophy concerned with principles related to human behavior, specifically the rightness and wrongness of such acts, as well as the goodness and badness of the intentions and purposes of certain actions. It is then subdivided into two sections:

Ethics based on axiology: an area of ethics concerned with the relative goodness or worth of the motivations and outcomes of any behavior.

DOI: 10.1201/9781003280989-1

1

Deontological ethics: a subset of ethics concerned with right conduct and the essence of duties, regardless of the goodness or worth of motivations or the desirableness of the ends of every act.

1.2 ETHICS IN BUSINESS WORLD

Market ethics is that type of implemented ethics or professional ethics that governs ethical principles and morals in the workplace. Philosophical ethics are reflected in business ethics. One of them is to figure out what a company's fundamental goals are. If a company's goal is to increase shareholder value while minimizing benefit to others, it is violating its duty. Corporate NIT'S are constitutionally regarded as a corporate parson of rights and responsibilities. The Importance of Ethics in the Corporate World: as business ethics is used as a guiding philosophy, employees make smarter choices in less time, which raises effectiveness and overall employee morale. The whole organization wins as employees carry out their duties with integrity and honesty (Figure 1.1).

A system of spiritual and ethical principles that govern a company's beliefs, actions, and choices, as well as the values, behaviors, and decisions of its workers, is referred to as business ethics. Environmental regulations, minimum wage legislation, and bans on stock trading and collusion are all instances of the government setting minimum business ethics standards. Company ethics has developed through time, and the different areas of ethics are important to every business.

Leadership for integrity

The executive committee establishes the tone for how the business operates on a daily basis. When a company's dominant management philosophy is focused on ethical values and behavior, leaders will set a precedent for their employees and encourage them to make decisions that favor both

FIGURE 1.1 Ethics in business world.

them and the company as a whole. Long-term advantages of establishing and maintaining a positive identity in society include the capacity to attract and retain highly qualified workers and the ability to create and maintain a good image in society. Running a business responsibly from the top down improves executive team ties, which contributes to the company's overall stability.

Employee integrity

When it comes to managing a company ethically, employees follow in the footsteps of managers. With corporate principles as a driving philosophy, employees make more decisions with less time, which improves performance and morale. The whole organization wins as employees carry out their duties with integrity and honesty. Employees who work for a business that maintains a high level of organizational honesty in all areas of management are more likely to perform well and remain loyal.

Ethics differ depending on the industry

Business ethics differ from industry to industry, as well as from nation to country. The nature of a company's operations has a significant effect on the regulatory issues it must address. When a customer's best option and their money do not coincide, the brokerage pays the maximum charge, for example, an ethical dilemma occurs for the brokerage. A broadcasting organization that produces children's television programming may have a moral imperative to uphold positive ideals and avoid using off-color material in its programming. Companies such as Amazon and Google, who perform the majority of their business online, are not subjected to environmental scrutiny. When it comes to protecting the identity and welfare of their clients, though, their ethics are scrutinized.

1.2.1 CORPORATE SOCIAL RESPONSIBILITY

It is a management concept that stresses the need of companies acting as responsible corporate citizens, not only by obeying the laws but also by conducting manufacturing and selling activities in a manner that does not damage the environment or deplete precious Earth capital. Some companies have started to act more socially conscious, partly because their executives wish to, and partly because they are afraid of environmentalist and customer pressure groups, as well as the media, and are concerned for their public profile. It is proposed that socially conscious action will pay dividends in the long run, even though it means sacrificing any short-term profits.

To be socially conscious, a company must first be accountable to itself and its shareholders. Companies that use CSR systems have often progressed to the point where they will give back to society. As a consequence, CSR is basically a business strategy. Furthermore, the more well-known and competitive a firm is, the greater its responsibility to establish ethical standards for its colleagues, rivals, and customers.

Starbucks has long been lauded for its commitment to environmental and community welfare, as well as a strong sense of CSR. Starbucks claims that it has achieved all of its CSR objectives since its inception. According to the company's Global Social Impact Report for 2019, these accomplishments include achieving 99% ethically

sourced coffee, establishing a global network of producers, pioneering green building in its markets, donating millions of hours of community service, and launching
an innovative college curriculum for its partner/employees.

Starbucks plans to hire 10,000 immigrants by 2020 and beyond, as well as reduce
the environmental footprint of its cups and include its staff in environmental foresight.
There are several socially conscious businesses, such as Ben & Jerry's ice cream and
Ever lane, a clothes chain, have products that are recognized for their CSR systems.

There are some factors that have made immoral activity more likely. Greater globalization has developed a much more dynamic work world that encompasses a wide
range of cultures and economies, making it much more difficult to apply ethical
values faithfully. Companies in the United States, for example, have gotten bad press
for relocating businesses to third-world nations, where workers live in circumstances
that would be unacceptable in other developed countries. Organizations are finding it
difficult to sustain sales and earnings in today's recessionary economy. To maintain
profitability, some businesses are tempted to engage in immoral acts. The Peanut
Corporation of America, for example, is accused of shipping contaminated goods
from its Georgia factory, which resulted in an epidemic in 2008 that killed at least 8
people and sickened over 550 people in 43 states.

Importance of Business Ethics

There are at least five compelling causes for businesses to promote a job climate
where workers are empowered to make ethical business choices.

1.2.2   REASONS FOR GOOD BUSINESS ETHICS

1. Obtaining the community's goodwill
2. Establishing a dependable company
3. Promoting ethical corporate practices
4. Defending the company and its staff from legal prosecution
5. Avoiding negative attention

Obtaining the community's goodwill

Organizations have certain basic societal obligations. These obligations
are sometimes stated in a written declaration of their company's values or
convictions. Many companies start or fund socially conscious program,
such as donating to charitable causes and nonprofit organizations, giving
incentives to staff who go beyond and above legal standards, and allocating
money to initiatives that are more socially desirable than lucrative. As a
result, many businesses start or endorse socially conscious initiatives, such
as donating to charity organizations and nonprofit entities. In 2011, IBM
workers in 120 countries volunteered 3.2 million hours of voluntary service.

Creating an organization that operates consistently

Values are developed and adhered to by organizations in order to establish a corporate ethos and identify a cohesive commitment to meeting
the interests of their stakeholders, who include owners, staff, consumers,
vendors, and the environment. Employees will realize what is required of
them and will be able to use the organization's principles to guide their

decision-making. Despite the fact that each company's belief structure is unique, all of them share the following principles:

- Operate with honesty and integrity, according to organizational values
- Operate in accordance with ethical conduct standards, both in words and deeds
- Treat colleagues, clients, and consumers with respect
- Strive to be the best at what matters most to the company
- Value diversity
- Foster good business practices

  Companies that provide excellent service retain their customers instead of losing them to competitors. Employee productivity and unemployment rates are lower in companies that establish and retain good employee relationships. Working with businesses that act in a fair and ethical way is also a concern for suppliers and other business partners. Bad ethics, on the other hand, will lead to poor business outcomes.

- Defend the company and its employees from legal action

  Establishing appropriate ethical and management systems is one way to discourage and detect workplace discrimination. Indeed, the Department of Justice published parole guidelines in 1991 that say that accused executives should be treated more leniently if their businesses have ethics policies. When an agency has developed an integrity compliance scheme that cooperates with police, fines for disciplinary offences will be reduced by up to 80%.

- Keep negative press at bay

  The valuation of a company's stock, how investors perceive its goods and services, the level of regulatory scrutiny it enjoys, and the extent of funding and assistance it receives from its corporate partners are all influenced by its public image. As a result, many businesses are inspired to develop a robust ethical policy in order to escape bad attention. Customers, corporate associates, lenders, consumer groups, financial firms, and regulatory authorities are more likely to see a company positively if it is considered to be acting ethically.

1.3 ETHICAL CONSIDERATION IN DECISION MAKING

Since they include inherent contradictions between a collection of economic and self-interest considerations and an opposing set of ethical, social considerations, and legal considerations, most important decisions in institutions are not only complicated but may be called dilemmas. Different solutions are favored by these conflicting factors. Although coming up with innovative solutions will help to minimize friction, some stress is almost always present. Some argue that organizations and management can only accept economic (organizational self-interest) considerations to escape the more complex challenge of balancing these competing variables. Others complicate by exaggerating an organization's or managers social responsibility (Figure 1.2).

What is "right" and "wrong," "just" and "unfair" in judgments and acts that affect others is the subject of ethics. Our ethical system is based on the principles that we

FIGURE 1.2   Steps involved in ethical decision making.

hold and consider essential. Values are a collection of moral values that guide our decisions and relationships on what is "right," "desirable," "just," and "of value." We test decisions and behaviors using certain criteria (values). For a variety of causes, ethical choices are almost always difficult. Most ethical choices involve: (a) several options; (b) long-term consequences; (c) unpredictable outcomes; (d) outcomes that combine different fiscal, legal, and social advantages and costs; and (e) personal ramifications. Organizational ethics decisions are not black-and-white choices; they require nuanced considerations weighing economic and self-interest benefits and costs against different legal, ethical, and social benefits and costs.

The following are four things that can help you figure out the legal aspects of a case:

- Objectives (considering multiple goals and their compatibility, plus constituent priorities)
- Techniques (including constituent acceptability, goal-satisfaction efficacy, and whether a proposed approach is essential, incidental, or extraneous to achieving the goals)
- Motivations (hidden or known, shared or selfish, and implicit value orientations)
- Resultant consequences (consideration of different time periods, impacts on constituents, and impacts on other partners and agencies)

1.3.1   DEVELOP A PROBLEM STATEMENT

A problem statement is a succinct, straightforward explanation of the problem that must be solved. A strong issue statement responds to the following questions: What do people notice that makes them believe there is a problem? Who are the people who are specifically affected by the problem? Is there someone else who has been impacted? How often does it happen? What is the problem's impact? What is the severity of the problem? The most important move in the decision-making process is to create an issue statement. It is pointless to continue without a simple statement of the issue or the decision to be taken. Obviously, the decision would not fix the problem if the problem is specified incorrectly.

1.3.2   IDENTIFY ALTERNATIVES

It is optimal to engage the support of others, including stakeholders, at this point of decision-making to find multiple alternative solutions to the issue. When you brainstorm with only one other person, the chances of identifying a wide variety of options and deciding the best answer are reduced. However, there are occasions when it is wrong to do so. Participate in the solution to a dilemma that you are unable to discuss. Provide the truth, not your view, when giving details on the issue to be solved, so you don't persuade others to consider your solution. Try not to be dismissive of proposals during any brainstorming session, as harsh feedback can seem to "shut down" the community and cause the flow of ideas to dry up. Simply jot down the thoughts as they come to you.

1.3.3   EVALUATE AND CHOOSE AN ALTERNATIVE

If a range of alternatives has been developed, the group aims to test them using a variety of metrics, including efficacy in solving the problem, the level of risk associated with each solution, cost, and implementation time. A solution that sounds appealing but is impractical would not help fix the dilemma. Weigh the different rules, standards, and principles that could apply as part of the selection process. You don't want to break a rule that might result in a fine or detention for you or anyone.

1.3.4   IMPLEMENT DECISION

If a solution has been chosen, it must be executed in a timely, reliable, and effective manner. Since people oppose transition, this is far better said than done. In reality, the more significant the transition, the greater the resistance. Communication is crucial in assisting citizens in accepting transition. The following questions must be answered by someone that the stakeholders support and respect: Why are we doing this? What's wrong with the way we're doing it now? For you, what are the advantages of the new method? A transformation strategy must be described in order to clarify to people how they can transition from the old to the new way of doing things. It is critical that the transformation be seen as simple and painless.

1.3.5   EVALUATE RESULTS

Track the outcome to see whether the desired consequence was accomplished, and observe the effects on the company and different stakeholders since the problem solution has been applied. Were any of the performance standards met? Did any negative consequences happen? This assessment will mean that further work is required. If this is the case, go back to the problem creation stage, make the relevant changes to the problem statement, and repeat the procedure.

1.3.5.1 The Ethical System's Nature and Problems

Moral values are established by an Eternal Law, which is expressed in scripture or manifested in nature, and then understood by religious leaders or philosophers. All should behave in compliance with the interpretation, according to the assumption.

Issue: There are various origins of the law and meanings of the law within each, but no means for choosing among them other than human rationality, which is motivated by both circumstance and self-interests. Human reasoning requires an absolute theory or concept as the foundation for decision-making, without an agreement: this completes a loop and returns us to the beginning.

Ethical egoism: Focuses on the individual's point of view. Individuals should pursue their own self-interests and behave to promote the greater balance of good over evil for themselves, according to this philosophy. In a free enterprise environment, ethical egoism may be seen as the default standard for companies. Some have viewed economist Milton Friedman's assertion that a company's sole duty is to maximize earnings as an example of the egoist viewpoint. Our own market-oriented economy is based on the assumption that people will obey this ethic while making purchases and other personal choices.

Issue: It could be argued that this is more closely a description of a "survival of the fittest" society than an ethical theory. Some argue that ethical egoism is a contradiction in terms; that is, there is no need to remind individuals they should behave in their own self-interest if they always do so. Since ethical egoism has no means of deciding conflicting arguments as they emerge, and no way of deciding whether one interest is more relevant than another, this perspective is very barren in terms of how mutual collaboration and competitiveness can be achieved. Furthermore, no egoist is in a position to provide guidance or render moral decisions on other people's behavior.

Utilitarianism: Moral expectations are extended to the result of an event or judgment (teleological theory) by all people influenced by the action, not only for them. The idea is that everybody should behave in the best interests of the greatest number of individuals (i.e., the greatest net social value to humanity, and the "greatest good for the greatest number," or maximization of the social gain function) (i.e., the highest net social benefit to society, and the "greatest good for the greatest number" or maximize the social benefit function). An action is "correct" if and only if it results in larger net gains for society than any other action conceivable in the circumstances. When utilizing this method, it's important to think of both positive and negative costs/outcomes, as well as nonmonetary benefits including wellbeing and friendships.

This system has a number of flaws:

a. Truth, health, peace, freedom, and enjoyment are examples of "goods" that utilitarian's think should be maximized.

    b. It says nothing about how the advantages are dispersed (e.g., it wouldn't disapprove of a slave-owning society if the overall amount of "goodness" was greater than an egalitarian society); and it says nothing about how the benefits are divided.

    c. It seems to be devoid of any sense of fairness or rights. Immoral actions may be "justified" if they help the majority, even if they come at an unacceptably high cost or cause damage to the minority. An extra concept or value is needed to balance the benefit–cost equation. There is no uniform technique for balancing the majority's advantages with the minority's sacrifices.

Universalism: Since the consequences are so indefinite and unclear at the moment the decision to act is taken, moral principles are related to the purpose of an event or decision. The idea is that everyone should take steps to ensure that others can make better choices under similar situations. This is a deontological (obligations or duties) approach and is the polar opposite of teleological philosophy. Universalism's first obligation is to regard people as ends rather than means.

Issue: It is difficult to objectively express and test one's behavior principle; for example, it might be possible to describe "immoral" acts in a particular manner. This ethical system does not have a basis for addressing disputes between duties (for example, a dispute between one's responsibility to one's business and one's responsibility to society as a whole). People who are vulnerable to self-deception or self-importance will "justify" immoral actions, and there is no scale to judge between "should"— there are no goals or degrees. (e.g., one person may prefer absolute rule and order, with no opposition to the government outside of formal elections, whereas another may prefer greater personal liberty.) To further refine the Categorical Imperative idea, an additional theory or value is required. It often assumes we aren't too clumsy or ineffective; otherwise, the system will fail!

Enlightened individuals self-interest: This is a system that combines utilitarianism and egoism. It can be described as "self-interest as a rational person understands it." If one takes a long-term view and considers one's own self-interests, an individual's self-interest and society's interests are (should be) close, according to ESI. True ethics recognizes that his or her own long-term needs and those of humanity are very similar. Using this context, one may ask: How would I rate this action if I were on my deathbed?

- Represents (or necessitates) an ideal (or ideal people) that is well above any of us. It is susceptible to rationalization and offers no support in practical implementations.

Interdependence ethics: In a culture based on a person's values, interdependence relationships between people are critical in deciding what is ethical. Personal responsibilities are emphasized over their rights. Two virtues encompass these responsibilities. Fair parties will still be able to cooperate in this arrangement, and each side is obliged and attempt and offer what the other side requires to achieve and accomplish its objectives. Any other kind of conduct is untrustworthy, exploitative, divisive, and immoral.

- This is indeed a representation of an ideal. It is mostly concerned with small-scale interpersonal experiences and is impossible to adapt to larger or more general contexts.

Distributive justice: The primacy of a single principle, justice, underpins moral values. All should work to ensure a more equal distribution of benefits for this fosters individual self-esteem, which is necessary for social participation.

Issue: Recognition of (a) justice as the first and foremost virtue of social institutions, and (b) the proposition that social cooperation is the foundation for all economic and social gains is required for the primacy of the value of justice (an equitable distribution of benefits ensures social cooperation). To different individuals, "justice" or "just distribution" means different things, such as equally, according to need, effort, contribution, competence, and so on. Individual efforts are overlooked.

Personal liberty: Morality is founded on the supremacy of a single principle, liberty. All should work to ensure that people have more freedom of choice, as this encourages business exchange, which is critical for social productivity. Libertarians advocate for this scheme.

- Recognition of (a) liberty as the prime value, the first condition of civilization, and (b) the proposition that a capitalist mode of commerce guarantees social efficiency is needed for the primacy of the value of liberty. This scheme is founded on a very strict conception of liberty, which is limited to the negative right to be free of others' interference; there could also be a positive right to partake of any of the privileges that some enjoy. The following are the four common paths to ethical decision-making:
  1. Virtue ethics approach
  2. Utilitarian approach
  3. Fairness approach
  4. Social good approach
- Approach based on virtue ethics

If you're dealing about your everyday life in a society, the virtue ethics guide to decision making reflects on how you can act and care about relationships. It does not describe an ethical decision-making formula, although it means that when confronted with a complicated ethical dilemma, people do what they are more comfortable with or what they believe a person they respect will do. It is assumed that people are motivated by their virtues in making the "right" decision. A practitioner of virtue ethics insists that doing the right thing is more effective than upholding a series of values and laws, and that moral deeds can be done out of habit rather than introspection. By equating the virtues of a good businessperson with that of a good person, virtue ethics should be extended to the business world. However, businesspeople face circumstances that are unique to the industry, so their ethics need to be tailored accordingly. The virtue ethics approach has the drawback of not including much of a roadmap for policy. There is no objective concept of virtue; it is determined by the circumstances—you figure it out when you go. For example, courage is a great virtue in certain situations, but it may still be stupid in other situations.

The best thing to do in a given case is determined by the society you're in and the societal norms that apply.

- A utilitarian perspective

   The utilitarian approach to rational decision-making says that you should choose the behavior or strategy that has the best net consequences for all people who are affected directly or indirectly. By weighing the needs of all involved individuals, the aim is to find the single greatest good. The economic concept of value and the application of cost-benefit analysis in industry are easily reconciled with utilitarianism. Business leaders, politicians, and scientists weigh the benefits and drawbacks of projects while deciding whether to invest in a new plant in a foreign market, adopt a new rule, or approve a new prescription medication. This method is complicated by the difficulty, if not impossibility, of calculating and comparing the prices of such gains and costs. How do you put a price on a human life or a pristine wildlife habitat? It can also be impossible to estimate the full extent of a decision's gains and drawbacks.

- Aim for fairness

   The justice approach reflects on how decisions and regulations share gains and responsibilities with those who are impacted by the decision in an equitable manner. The approach's driving philosophy is to consider all equally. However, psychological animosity against a single party may affect decisions taken using this approach, and policy makers may not even be aware of their prejudice. If the aim of a policy or intervention is to favor a certain group, other impacted groups will find the decision to be unjust.

- Aim for the common good

   A view of society as a group where members work together to accomplish a shared set of principles and priorities underpins the common good approach to decision-making. This methodology is used in decisions and strategies that aim to enforce social processes.

1.4 DIFFERENCE BETWEEN MORALS, ETHICS, AND LAWS

Morals refer to one's religious views on what is right and wrong, whereas ethics refers to the rules or codes of conduct that a society (nation, institution, or profession) expects of an individual. For example, the principles of the legal profession require lawyers to protect an innocent defendant to the best of their abilities, particularly though they are well aware that the client has committed the most terrible and morally repugnant crime conceivable (Figure 1.3).

Law is a set of laws that determines what we are permitted to do and what we are not permitted to do. A set of agencies is in charge of enforcing laws (police, courts, law-making bodies). Actions that comply with the law are known as legal deeds.

A moral action backs up what an individual believes is the right thing to do.

Laws will declare an act lawful, even though certain people believe it is immoral—for example, abortion.

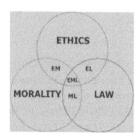

FIGURE 1.3 Differences between morals, ethics, and laws.

- Integrity

     Ethics is a field of philosophy that attempts to answer the question, "What should I do?" It's a self-reflective process in which people make decisions based on their values, morality, and meaning rather than habit, societal conventions, or self-interest. Our ideas, goals, and reason provide us with a sense of what is good, right, and essential in our life. They act as a reference point for all of the choices that are accessible to us. According to this idea, an ethical judgment is one that is based on reflection on the issues that matter to us and is consistent with our beliefs. While each person has the capacity to concentrate on and find their own sense of what is good, right, and essential, throughout history, varied groups have come together around various sets of ideas, goals, and values.

- Righteousness

     Many individuals benefit much from morality. Not everyone has the time or the training to figure out what kind of life they want to live, taking into consideration all of the many ideas, values, and aspirations that exist. It's advantageous for them to have a well-organized, dependable account that they can duplicate in their everyday life.

     Morality is typically passed down from one's family, community, or culture; it's rare for someone to go out and find the morality that best fits their own views. The majority of the time, the procedure is unconsciously carried out. There's a dilemma here: if we inherit a ready-made answer to the issue of how to live, we'll adapt it to our life without ever assessing whether it's enough. We may spend our entire lives adhering to a moral code that, if given the opportunity to consider it, we would oppose in part or in whole.

- Legislation

     In this instance, the rule is different. It is not a strict morality, but it seeks to provide a private space where individuals may act according to their own ethical beliefs or morals, at least in democratic nations. Instead, the law seeks to provide a basic, enforced code of behavior that is necessary for a community to exist and everyone to be treated equally. Enforcing laws is the responsibility of a group of authorities.

     As a consequence, the statute's reach extends beyond ethics and morality. Although the rule is agnostic on certain topics, ethics and morality have a lot to say. The law won't assist you in deciding whether to notify your competitor that their new client has a history of not paying their bills, but our views about what's good and right will.

It's tempting to conflate law with ethics, believing that as long as we follow the law, we're being ethical. On two counts, this is incorrect. First, the legislation establishes a fundamental code of conduct that is required for our social institutions to continue to operate. It safeguards fundamental consumer protections, for example. However, in certain cases, doing the right thing in resolving a consumer dispute might necessitate us going beyond our legal obligations.

Second, there may be times when adhering to the law compels individuals to act in ways that are incompatible with our values or morals. A doctor may feel forced to perform an unlawful procedure, or a government official may believe it is their duty to disclose sensitive material to the press. Some thinkers argue that a person's morality, rather than the legislation, binds them, suggesting that obeying the text of the law isn't an adequate substitute for ethical reflection.

The distinction between ethics and law is clear, though they can (and do) intersect, as in all laws and regulations. The distinction between ethics and justice is much more pronounced. To act within the principles that they have promised to defend, a prosecutor will not always be able to offer justice. When a prosecutor suspects their client is guilty of the crime for which they are defending them, anything will happen. The distinction between ethics and morality is obvious in this case, but you must protect your client regardless of your objections in order to fulfill your obligation as a defense lawyer. However, ethics are not always the rule, and in most occupations, you can speak out if you believe you have a moral duty to report ethical or legal breaches in the workplace. In the corporate world, ethical questions abound. It is your responsibility to inform your employer or another superior within the company of the situation in these cases.

1.5 NECESSITY OF ETHICS IN INFORMATION TECHNOLOGY

For those who have access to it, knowledge is a source of power and, increasingly, the gateway to success. As a result, the implementation of information systems is intertwined with social and political affairs, making ethical considerations of how information is handled much more important. Electronic devices have infiltrated all aspects of governance, work, and personal life to the point that even those without access to them have had a huge impact on how they operate. New ethical and legal decisions are required to strike a balance between everyone's interests and rights (Lynch, 1994).

Ethics fills the void, while court rulings lag behind technological advancements.

Court rulings in the face of technological change, as in most emerging technological areas, are missing. Ethics will help to fill the void in people's understanding of how to negotiate the use of electronic content. The accompanying notes outline the main ethical topics being discussed. These notes can be read in accordance with the legal issues of electronic information systems and rules to rule on some parts of these issues have been made (Lynch, 1994).

Electronic knowledge system ethical issues

Specific moral decisions in contrast to other cultural expectations of ethical behavior and guidelines for members of the industry are referred to as ethics. The regulation of electronic records and access to information, privacy and data manipulation, and international implications are all broad concerns. All of these apply to computer networks, electronic databases, and geographic information systems in particular. Specific issues for each of the three, on the other hand, necessitate slightly different ethical decisions. Above, we'll look at networks, electronic information systems in general, and geographic information systems in particular (Lynch et al., 1994).

Privacy policy for computer ethics

Aside from apparent illegal behavior, less obvious types of machine activity may pose ethical concerns. Employees' use of computers for personal purposes, for example, has sparked heated controversy, but no concrete solutions have emerged that can be extended to all organizations. Most workers who use laptops have an email address and review it often at work. In general, the company's important internal messages are often delivered by email. Employees, on the other hand, will be given a personal email address and may use the business computer to send and receive personal messages.

New technologies permit not only the checking of email, but even other Internet practices such as message boards, text, and even web browsing. While businesses will want to ensure that their workers spend their time with the company, track network traffic, and catch ethical lapses, this is difficult to do because the motivations for using a website cannot be decided merely by recognizing that someone visited it. This issue will continue to divert attention away from the business. Given the meaning of a mode of contact or a visit to a website that could be unfamiliar outside of the monitor, there is a considerable risk of confusion, misinterpretation, and manipulation of the data obtained. In the late 1990s and early 2000s, the controversy over secrecy and regulation of corporate email and other electronic activity was one of the most well publicized legal issues. Although businesses contend that their own management mechanisms are required to make appropriate and profitable use of company time, the moral right to privacy has been consistently asserted.

Importance of ethics in the field of information and technology

1. With the proliferation of the Internet, the opportunity to collect and retain large volumes of personal information, and a greater dependence on computer systems, the possibility of information technology being used unethically has risen.
2. Examples of ethical information technology use that arouse public concern:
   - The email and Internet use of many employees is tracked when they are at work, while managers want to strike a balance between their ability to handle vital self-direction and their need to control productivity.
   - Millions of people have downloaded free music and movies, ostensibly in breach of copyright rules, at great cost to the copyright owners.
   - Organizations use unsolicited email (spam) to reach millions of users around the world as a low-cost communications strategy.

- Hackers gain access to the files of financial and retail institutions in order to harvest consumer records, which they then use to conduct identity fraud by creating new identities and making payments on the credit cards of unsuspecting suspects.
- Students from all over the world have been found plagiarizing content for their term papers by copying material from the Internet.
- To log travelers' online transactions and events, websites place cookies or spyware on their hard drives.

1.6   ETHICS FOR IT PROFESSIONALS

The information technology industry is said to be the world's fastest developing and most diverse industry. Computers have become an integral part of everyone's lives, and they are used in a variety of fields including education, networking, industry, entertainment, building, medicine, and security. Information technology is used to create communications campaigns for entrepreneurs and businesspeople, accounting software for financial firms, database creation for effective networking, resource control, and customer service, as well as the development of technologies and equipment for agriculture, security, medicine, engineering, and other fields. In layman's words, an IT specialist is someone who works in the field of information technology.

The following characteristics are needed for a good tech career:

1. Patience is a key characteristic; the IT technical world is enormous, and IT professional resources necessitate continuous analysis, trial, and error attempts to arrive at a specific solution. As a result, one must be patient with the individuals and equipment with which one operates. Before coming to any conclusions, the dilemma should be carefully examined, and hasty assumptions and assessments should be avoided.
2. He/she should be able to communicate effectively. When engaging with various types of audiences, he/she should be understandable, clear, and concise. It is important to have expertise and experience dealing with various styles of individuals using various communication techniques.
3. The area of IT technical definition is constantly changing and evolving. One must be able to adapt quickly and have an ability to explore new technology and applications. It is important to keep one informed.
4. An IT specialist should be able to do several tasks at the same time. For example, when restoring a device, he/she will need to address both hardware and software issues. Around the same time, he/she could have learnt of a new version of software or a computer that may help him/her solve the problem more quickly.
5. Problem-solving ability is a desirable trait. Often, especially in the fields of networking, software, and program creation, one can encounter problems with which he/she is unfamiliar. He/she will have to use his/her own problem-solving philosophies and methods, as well as the scientific skills he/she has gained from experience and study, at that stage.

6. A tech professional should be excited about his or her job, love what he or she does, and be willing to learn more about the industry and be able to apply what he or she has learned.

1.6.1 PROFESSIONALS

A career is a vocation that necessitates advanced experience and, in many cases, extensive academic training. An individual "working in a professional career capacity" is described by the US Code of Federal Regulations as someone who meets the following four criteria:

- One's primary responsibilities include doing work that necessitates advanced expertise in an area of science or engineering typically gained through a lengthy course of specialized scientific instruction and research or work.
- One's instruction, research, or practice is unique and imaginative in nature in a recognized area of artistic achievement, the outcome of which is solely dependent on the employee's professional invention, imagination, or ability.
- When it comes to professional success, one's role necessitates the consistent application of discretion and judgment.
- One's practice is mostly intellectual and diverse in nature and the performance career or outcome cannot be standardized over a specific time frame.

In other words, specialized qualifications and expertise are required of practitioners such as physicians, attorneys, and accountants; they must exercise independence. In the course of their job, they must exercise discretion and judgment, and their work cannot be structured.

Many individuals, especially professionals, are expected to contribute to society and participate in a program of lifelong learning (both formal and informal), to stay current in their career and to help other professionals develop. Furthermore, certain technical positions come with unique rights and obligations.

- Doctors, for example, administer medications, operate on patients, and seek sensitive medical records while retaining doctor–patient confidentiality.

IT Professionals are those who work in the field of information technology.

1. Many workers in the business, such as communications analysts, financial advisers, and IT specialists, have duties, backgrounds, and experience that qualify them as professionals.
2. IT professionals include engineers, systems administrators, and software industry knowledge officers, to name a few (CIOs).
3. One might contend, though, that not all IT position necessitates "advanced experience of an area of science or learning customarily gained by a career lengthy course of professional intellectual teaching and study," to borrow from the United States Code of Federal Regulations.IT employees are not

accepted as practitioners in the official sense and they are not approved by the state or federal governments.
4. The difference between career and specialist is relevant in malpractice cases, as many courts have found that professional IT employees are not responsible for malpractice because they do not follow the legal description of a professional.

1.6.2 PROFESSIONAL CODE OF ETHICS

A professional code of ethics outlines the ideals and standards that are fundamental to a certain occupational group's practice. Most occupations have a code of ethics that regulates how they conduct themselves. Doctors, for example, follow various variations of the 2000-year-old Hippocratic Oath, which medical schools give to their graduate students as an affirmation (Figure 1.4). Most technical organization's codes of ethics are divided into two sections:

• The first specifies the organization's goals
• Second usually lays down the guidelines and standards that all members are supposed to follow.

Laws should not serve as a comprehensive roadmap to ethical conduct. But, adhering to a professional code of ethics will negate a number of advantages for individuals, professions, and society as a whole:

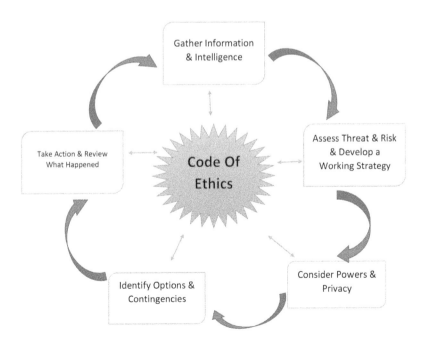

FIGURE 1.4 Professional code of ethics.

- Ethical decision-making—Adherence to a professional code of ethics requires that professionals make ethical decisions based on a shared collection of core principles and convictions.
- High ethical principles and practice—Professionals who follow a code of ethics are reminded of the responsibilities and roles they may be tempted to breach in order to meet the demands of day-to-day business. To instruct practitioners in their relationships with others, the code also specifies permissible and inappropriate practices.
- Public interest and confidence—public trust is based on the presumption that a practitioner can act ethically. People often rely on a professional's honesty and sound judgment to say the truth, refrain from providing self-serving advice, and provide warnings about the possible negative consequences of their decisions. As a result, following a code of ethics increases faith and respect for practitioners and their field.
- Self-evaluation benchmark—a code of ethics serves as a self-evaluation benchmark for professionals. The coding may also be used by the professional's peers for praise or criticism.

Seven trends are reshaping the professional services industry:

1. Client sophistication
2. Governance
3. Interconnectedness
4. Discretion
5. Modularization
6. Globalization
7. Commoditization

1.6.2.1   Client Sophistication

Clients are more aware of what they need from service providers, are more inclined to go outside their own organization for the appropriate benefits, and are more capable of negotiating hard to receive the finest services at the lowest feasible cost.

1.6.2.2   Governance

Major scandals, as well as stricter legislation enforced to prevent further scandals (e.g., Sarbanes-Oxley), have created an atmosphere in which clients and service providers have less confidence and more oversight.

1.6.2.3   Interconnectedness

Clients and service providers have established their business ties on the expectation of a positive outcome that is being able to communicate easily and quickly anywhere in the world through electronic teleconferences, audio conferences, email, and wireless devices.

1.6.2.4   Discretion

Clients want to monitor and manage work-in-progress in real time. Clients are no longer able to wait until the final product is complete before providing input and comments.

1.6.2.5   Modularization

Clients should deconstruct their business processes into their basic steps and determine which ones they can handle directly and which ones they can outsource to service providers.

1.6.2.6   Globalization

Clients will compare and choose from a wide range of service providers all around the world, rendering the service provider market highly competitive.

1.6.2.7   Commoditization

Low-end service delivery (e.g., personnel augmentation to complete a project) is seen by clients as a commodity service when the main criteria for choosing a service provider is pricing. Clients aim to establish partnerships with their service providers in order to provide high-end offerings (e.g., the implementation of an IT business plan).

1.7   PROFESSIONAL RELATIONSHIPS

Organizational, corporate, organization, and market partnerships all share one thing in common: coordination and work relationships. There is no pillar on which an organization can stand without working relationships. Also for small companies owned by sole proprietors or owners, maintaining a positive interaction with dealers and consumers is critical to their success. There will be little advancement or achievement without a community of individuals who devote their talent and abilities to the creation of an organization (Figure 1.5).

1.7.1   Relationship between IT Workers and Employers

IT employees and employers have a complex and important relationship that takes constant commitment from both sides to maintain. Before accepting a job offer, an IT worker and his or her boss usually negotiate on the fundamentals of their relationship. Job title, general performance standards, basic work obligations, drug-testing regulations, dress code, workplace, pay, work hours, and company benefits are examples of these concerns. Many such topics are covered in the company's rules and handbook of procedures or, if one exists, in the company's code of ethics. These concerns include the security of business secrets, vacation policies, time off for a wedding or a personal illness, tuition repayment, and utilization of company infrastructure, such as machines and networks. Although end users are often blamed for using unauthorized versions of proprietary applications, software theft in the workplace can often be traced back to IT staff members, who either cause it to happen or deliberately participate in it to save money on IT. The Business Software Alliance (BSA) is a

FIGURE 1.5   IT professionals to manage a variety of professional relationships.

trade association that represents the leading software and hardware companies in the world. Its aim is to prevent unlicensed copies of software created by its members. Thousands of cases of software theft are prosecuted each year by more than 100 BSA attorneys and prosecutors. Trade confidentiality is another place where IT staff and their employers can face difficulties. A trade secret is information that a corporation has taken great care to keep private and that is usually unavailable to the general public. It reflects something of economic worth that took time or money to produce and has a degree of innovation or uniqueness.

The development of contemporary software code, hardware prototypes, business strategies, and the design of a computer application's user interface, and production techniques are also examples of trade secrets. Whistle blowing is another problem that can cause discord between employers and IT staff. Whistle blowing is an employee's attempt to draw attention to a company's incompetent, immoral, dishonest, violent, or hazardous act that jeopardizes the public's interest. Because of their experience or status inside the infringing agency, whistleblowers also have unique knowledge. A chip-processing industry employee, for example, should be aware that the chemical procedure used to manufacture the chips is hazardous to all workers and the general public. A diligent person will bring the issue to management's notice and attempt to resolve it by collaborating with suitable organization tools.

1.7.2   RELATIONSHIP BETWEEN IT WORKERS AND CLIENTS

IT employee also offers services to customers who are either "internal" or "external" to the company. Each participant in an IT worker–client partnership decides to give something of value to the other. In general, a computer technician provides hardware, software, or services at a fixed fee and within a certain time frame. For instance, an IT employee might agree to put in place a new accounts payable software package that fits the client's needs. The customer pays you, gives you access to important connections, and maybe even gives you a place to live. This connection is mostly recorded through contractual language, which specifies who does what, when work starts, how long it will take, and how much the customer will pay, among other things. About the fact that IT staff and their customers frequently have vastly different levels of technological competence, the two sides must collaborate to be competitive. Fraud is the deceit or trickery used to procure goods, services, or property. Fraudulent misrepresentation happens when someone intentionally tries to persuade another person to believe and act on a lie. A misstatement or incomplete statement of a factual fact is referred to as misrepresentation. If the misrepresentation leads to the other side entering into a deal, the other party will be able to void the contract or claim damages compensation.

When one side fails to satisfy the terms of a contract, it is called a breach of contract. Furthermore, a substantive violation of contract happens when one side fails to fulfill such explicit or implicit commitments, causing the contract's meaning to be harmed or destroyed. Since there is no reasonable distinction between a minor and a substantive violation, a case-by-case decision is made. "When a substantive breach of contract" occurs, the nonbreaching party has following two options:

1. cancel the contract, demand restitution of any costs paid to the breaching party under the contract, and be released from any further contract performance; or
2. Consider the contract as being in effect and sue the breaching party for damages.

1.7.3 RELATIONSHIP BETWEEN IT WORKERS AND SUPPLIERS

IT professionals collaborate for a variety of infrastructure, applications, and service providers. Most IT professionals recognize that cultivating a positive working relationship with vendors promotes the transfer of relevant information as well as the exchange of ideas. Such awareness may lead to new and cost-effective ways to use the supplier's goods and services that the IT worker could not have noticed before. Bribery is whether a company or government official is given money, property, or favors in exchange for a business advantage. A tech distribution agent who gives money to a competitor's employee in exchange for business is an apparent example.

A kickback or reward is a term used to describe this kind of bribe. When a reward is offered, the one who makes the deal commits a felony, and the person who accepts the bid commits bribery. Various jurisdictions have passed bribery laws, which have been used to invalidate bribe-related contracts on occasion but are seldom used to obtain felony prosecutions. Gifts are an important aspect of doing business in certain countries. In particular, in certain nations, failing to bring a gift to an initial business meeting is deemed impolite. In the United States, a donation could be a complimentary admission to a sports event from a staffing firm hoping to be added to the company's chosen vendors list. The most important distinction is that no gift can be concealed. If a donation is not announced, it could be called a bribe. As a result, most businesses expect all presents to be announced, with the exception of token gifts, to be refused. Some businesses have a policy of pooling employee donations, auctioning them off, and donating the proceeds to charity. When it comes to discriminating between bribery and donations, the donor's and recipient's perspectives will vary. The recipient may assume he or she has received a gift that does not bind him or her to the donor in any way, particularly if the gift was not in the form of cash. The donor's motives, on the other hand, may be very different.

1.7.4   RELATIONSHIPS BETWEEN IT WORKERS AND OTHER PROFESSIONALS

Professionals have a strong sense of commitment toward their colleagues. Members of a career also apply professional respect to one another, cooperating and supporting one another in their actions and decisions, according to the document. In the real world, IT staff often choose to either not exchange information with one another or keep all information to themselves. As a result, they are eager to assist one another in obtaining new positions but reluctant to publicly threaten one another. Professionals are often concerned with the perception of their field as a whole, and how it is viewed has an impact on how individual participants are regarded and handled.

As a result, practitioners owe each other and the professions code of ethics. Mentorship and development of potential members of the discipline will also be provided by experienced practitioners. Members of the IT profession can face a variety of ethical issues. Résumé inflation, which entails lying on a résumé and alleging expertise in a high-demand IT talent, is one of the most popular issues. Although exaggerating an IT worker's credentials can help him or her in the short term, it may harm the profession and the employee in the long run.

1.7.5   RELATIONSHIP BETWEEN IT WORKERS AND USERS

The word "IT customer" refers to the individual who uses a hardware or software device as opposed to the IT staff who design, mount, service, and help it. Understanding the end user's desires and skills, as well as creating an environment that encourages ethical action, tend to be the most critical points in this debate. You may do this by speaking with customers to learn what they need to know, and presenting them with helpful cheat sheets or hardware and software manuals. You can also make sure the compilation disks and image files are not stored in places where end users might take or copy them, to prevent the problem of unlicensed device installation. Locking down user accounts so they can't alter the configuration of a device may irritate certain users, but it also eliminates the risk of them downloading or deleting software without permission. The product must have corporate advantages or improve efficiency for IT consumers. IT professionals have a responsibility to consider a user's expectations and skills, then develop goods and services that better address those needs—within budget and time limits, of course. IT staff will have a critical role to play in creating an atmosphere that encourages users to act ethically.

Software piracy is discouraged, improper usage of corporate computer services is minimized, and inappropriate knowledge sharing is avoided in such an environment.

1.7.6 Relationship between IT Workers and Society

To protect the public, regulatory regulations define safety requirements for goods and services. This is a touch of an exaggeration. Some IT employees perform tasks that have an impact on those outside of their business and its clients, but this is the case rather than the norm. If you're one of the very few, you have the same responsibilities as any other person whose job has an effect on the general population. The public has confidence in those who represent them to look out for their best interests. Nevertheless, these rules aren't flawless, and they don't protect from any of a product's or process's harmful side effects. Professionals may frequently see the impact of their practice and take steps to mitigate future public risks. As a result, society requires practitioners of a discipline to have substantial gains while avoiding injury. Establishing and maintaining ethical practices that protect the public is one way to satisfy this expectation.

1.8 INFORMATION TECHNOLOGY PROFESSIONAL MALPRACTICES

The responsibility to protect individuals from undue injury or danger is known as the duty of treatment. Humans, for example, have a responsibility to protect their dogs from harming others and to drive carefully. Companies, likewise, must keep harmful chemicals out of air and water, manufacture clean materials, and ensure safe working practices for their workers. Courts use the fair person test to determine whether parties owe a duty of care by evaluating how an impartial, cautious, and attentive person may have behaved under the same circumstances. In the same way, defendants with specific experience or competence are held to a fair ethical level. Consider a fictional incompetence case in which an individual accidentally deleted millions of consumer documents from an Oracle database. Failure to behave in the manner a responsible person will act is a violation of the duty of care. A violation of duty may be an action, such as tossing a lit cigarette into a fireworks factory and triggering an accident, or a failure to intervene while a duty exists, such as a police officer failing to shield a person from an intruder. Professionals who fail to uphold their duty of responsibility are responsible for all injury caused by their incompetence. Professional malpractice is the term used to describe this type of liability.

IT technical responsibilities include:

* Recognize the user's requirements and capabilities.
* Provide a substance which lives depending on the needs of IT technical person.
* Determine the number of students who will enroll.
* Encourage high professional standards and behavior.

Organizational structure:

IT professional certification shows that you are an expert in the field of information technology.

1. A professional is someone who has a specific set of skills, experience, and talents in the application of a licensed system.
2. In comparison to lassoing, general valedictory is minimal.
3. There is no obligation to follow a code of ethics.
4. It may be used as a benchmark for learning a certain talent or expertise.

1.9   COMMON ETHICAL ISSUES FOR IT USERS

1.9.1   SOFTWARE PRIVACY

In a business environment, software piracy may often be traced back to IT experts, who can either encourage or actively participate in it. Corporate IT access practices and administration should enable consumers to report and contest cases of piracy.

1.9.2   INAPPROPRIATE USE OF COMPUTING RESOURCES

Many workers use their laptops to go to famous websites that have little to do with their careers, engage in chat rooms, scan pornography websites, and play video games. These tasks take up resources and decrease job production. Furthermore, behaviors such as viewing sexually graphic content, telling inappropriate jokes, and sending hate email could result in complaints and charges that an organization allowed a racist or sexual harassment-friendly work atmosphere. Frequent pornography users are often shot, although less serious offenders face administrative action.

1.9.3   INAPPROPRIATE SHARING OF INFORMATION

Any business has a huge volume of data that can be categorized as either private or confidential. A worker's private data include items such as wage information, attendance reports, health records, and performance reviews. Private data also contain details for consumers, such as payment card numbers, phone numbers, and home addresses. Confidential information include sales and marketing calendars, personnel projections, manufacturing processes, formulation formulas, tactical and strategic strategies, and research and development. Perhaps accidentally, an IT consumer who communicates this information with an unauthorized party has compromised someone's privacy or generated the possibility of business information falling into the hands of rivals.

When establishing an IT use policy, businesses may do one or more of the following things:

* Establishing guidelines for use of company software
  IT administrators at businesses must provide specific guidelines for the usage of personal computers and applications. Any firms work out

agreements with tech providers to include PCs and software such that IT staff can work from home. Other firms assist workers with acquiring lawful copies of all the software they need to be profitable, whether they operate in a workplace, on the road, or at home.

- Defining and limiting the appropriate use of IT resources

    Employees must be encouraged to respect company IT resources and utilize them to improve their work performance by developing, communicating, and enforcing written standards. Employees are allowed limited personal usage while being prohibited from accessing undesirable websites or sending rude or harassing messages through corporate email.

- Structuring information systems to protect data and information

    Organizations must establish procedures and standards to limit data access to just those employees who need it. Through a corporate network, sales managers, for example, may have complete access to sales and marketing datasets, but their access can be restricted to items over which they have responsibility. If they don't require access to things such as research and development results, formulation formulations, or personnel estimates, they can be refused access. It may also be set up to restrict those forms of attachments in email.

How to revolve ethical behavior prominently?

1. With employees what ethical behavior is expected?
2. Be consistent so that rules apply to everyone's workplace.
3. Let employees know how they should act before you.
    - Setting up and keeping a corporate firewall

        A firewall is a hardware or software interface that serves as a gateway between a company's network and the Internet while also restricting network access based on the company's Internet use rules. Through restricting links to individual undesirable webpages, the firewall will act as an effective barrier to illegal Web browsing. Unfortunately, the number of such pages is constantly increasing, making it impossible to ban any of them. Incoming email from specific Web addresses, businesses, or users will be effectively blocked by the firewall. It can also be set up to restrict those types of attachments in email.

How to prominently revolve ethical behavior

1. What is demanded of workers in terms of ethical behavior?
2. Maintain consistency such that the rules extend to all at work.
3. Explain to staff how they should behave in front of you.

In the workplace, ethical behavior is important.

The convergence of all people's eagerness is a sensor element in an ethical workplace in a society that offends features intellectual and creates confusion between classes.

1.10   CERTIFICATION

Certification means that, in the opinion of the certifying body, a practitioner has a certain range of talents, expertise, or ability. Unlike license, which is only applicable to individuals and is mandated by statute, registration can be applied to both people and goods and is usually optional. IT certifications may or may not contain a provision to follow a code of ethics, while licensing does. Certifications are offered by a variety of corporations and educational associations, and views about their merit are mixed. Many companies regard them as a barometer of mastery of a certain range of fundamental skills. Some hiring managers, on the other hand, are skeptical of certifications because they are no substitution for experience and do not guarantee that an individual can perform well on the job. Most IT professionals want to improve their talents, and qualification is a formal means to do so. Completing a credential offers strong appreciation for these individuals, as well as a plan to help them develop and succeed in their professions. Others see registration as just another way for commodity sellers to make more money with little value added.

1.10.1   VENDORS CERTIFICATION

Certain certifications boost a worker's wage and career prospects, which is essential for tightly identified roles within a larger position. It necessitates the passing of a written test, which usually entails certification and seldom involves the use of modern technologies: Microsoft-accredited technical developer SUN-certified associate. The government pays attention at the administrative level, for example. Engineers who provide public infrastructure services include public accountants, doctors, lawyers, and engineers. It encourages professionals to adhere to the IES code of ethics.

It therefore necessitates the adoption of a common code of ethics by IT professionals. IT professional licensing can boost today's extremely demanding IT system. Government licensing of IT practitioners is a technical problem. For IT professionals, there are both international and state licensing programs. There is no governing authority in charge of the acridity education scheme.

1.10.2   INDUSTRY ASSOCIATION CERTIFICATION

There are several business certifications available in a number of IT-related subject fields. The importance of these certifications vary widely based on where individuals are in their careers, what other certifications they have, and the essence of the IT work market. Individuals seeking certification must normally have the necessary qualifications and experience, sit for and pass an exam, and adhere to and follow a code of ethics defined by the institution offering the certification. They must pay a yearly fee, complete continuing education credits, and, in certain instances, pass a monthly renewal test to maintain their license. Industry association certifications often require more expertise and a wider outlook than manufacturer certifications; however, industry associations often fall behind in designing tests to include emerging technologies. IT certification is shifting away from strictly technological content and toward a wider mix of academic, industry, and behavioral competencies that are expected in today's challenging IT roles. This pattern can be seen in business organization certifications for diverse positions including project management and network security.

REVIEW QUESTIONS

1. Why business ethics is becoming increasingly important? Explain in detail.
2. What is ethics? Why it is important to act according to code of ethics.
3. Write various approaches for Ethical Decision Making.
4. What are the common ethical issues for IT users?
5. Explain ethics, moral, and law with suitable example.
6. Fostering good business ethics is important. Justify.
7. How decision making can be improved by ethical consideration?
8. Discuss the key ethical issues that can arise in each relationship an IT worker needs to manage.
9. Explain the actions taken by an organization while creating IT usage policy.
10. "Good ethics can mean good business" Justify the statement.
11. Discuss IT professional malpractices in detail.
12. "There is no requirement of ethics in IT or in any business." State whether the above statement is true or false? Justify your answer.
13. Explain the term "IT Professionals." What are the common ethical issues for IT users?

MULTIPLE CHOICE QUESTIONS (MCQs)

1. **The connection between ethics and the law may be summarized as**
   a. In reality, in every case, what is ethical is very near.
   b. No, there is a little overlap between ethics and the law.
   c. Close, since there is a lot of overlap between ethics and law, but they are not same.
   d. Not at all close, ethics and the law are virtually incompatible.
2. **Businesses have _____ throughout society, having the ability to give a _____ to society while _____ has the potential to do a great deal of damage to people, communities, and the environment.**
   a. Little power, Small contribution, Business malpractice
   b. Little power, Major contribution, Corporate social responsibility
   c. Corporate social responsibility, Huge power, Small contribution
   d. Huge power, Major contribution, Business malpractice
3. **IT Professional have to maintain relationship with**
   a. Client
   b. Supplier
   c. Other Professionals
   d. All of the Above

4. ____ is a major influence on the present and future status of corporate ethics, particularly in the area of ____, ____ and ____ related
issues.
   a. Globalization, Cultural, Legal, Accountability
   b. Nationalization, Cultural, Environmental, Accountability
   c. Privatization, Cultural, Legal, Accountability
   d. Globalization, Technological, Financial, Accountability

5. The moral principles, standards of behavior, or set of values that guide
a person's actions in the workplace is called
   a. Office place ethics
   b. Factory place ethics
   c. Behavioral ethics
   d. Workplace ethics

6. Values and ethics shape the
   a. Corporate unity
   b. Corporate discipline
   c. Corporate culture
   d. Corporate differences

7. Which of the following factors encourages good ethics in the workplace?
   a. Transparency
   b. Fair treatment to the employees of all levels
   c. Both (a) and (b)
   d. bribe

8. ____ is a self-replicating software that has the potential to corrupt data
and files on your computer.
   a. Freeware
   b. Virus
   c. Worm
   d. Piracy

9. How does a person acquire the ability to make moral decisions?
   1. By understanding his motivations
   2. By comprehending the ramifications of his actions
   3. By being discouraged and scared by the consequences imposed on him
      or her for his or her conduct
   4. By comprehending the methods used to carry out action
      a. 1, 2, and 4
      b. 1, 2, 3, and 4
      c. 1, 3, and 4
      d. 1, 2, and 3

10. A code of ethics serves which of the following purposes?
   1. It may indicate the members of a professional society's common com-
      mitment to uphold particular ethical norms and values.
   2. It may aid in the creation of an atmosphere where ethical conduct is the
      norm.
   3. In certain circumstances, it may be used as a guide or reminder.

4. For a company, a code may be a significant intellectual and educational legacy.

   Which of the following statements is true?
   a. 1, 2, 3, and 4
   b. 1, 2, and 4
   c. 1 and 3
   d. 1 and 2

ANSWERS TO MCQs

Q1: c, Q2: d, Q3: d, Q4: a, Q5: c, Q6: c, Q7: c, Q8: b, Q9: d, Q10: a

2 Computer and Internet Crime

LEARNING OBJECTIVES

- Examining the worms and viruses that influence Internet use and security.
- Recognizing critical computer crime concerns.
- Defining the importance of cybercrime laws.
- Investigating the effects of cyber terrorism.

2.1 INTRODUCTION TO CYBERCRIME

Cybercrime is a form of illegal activity that takes place over the Internet. The above is a broad description of cybercrime, and it is an unlawful Internet operation in which a device is used as a weapon, goal, or both. Illegal entry, illegal intrusion, and computer abuse are examples of cybercrime offenses against information technology (IT) infrastructure.

In a nutshell, cybercrime refers to any criminal operation carried out using modern technology. For instance, data theft is one of the most popular forms of cybercrime, but cybercrime often encompasses a broad variety of destructive behavior, such as cyber bullying or the distribution of worms or viruses. Cybercrime can be classified into two types: those that cause malicious harm and those that cause accidental harm. The most common type of harm is financial, although this is not always the case. Most people are aware of the phenomenon of cybercrime by now, but they might not be aware of the complete implications or expense. The most well-known example of cybercrime is hacking for the intent of obtaining financial or personal details, but it is far from the only one (Figure 2.1).

When cyber bullying poses a danger to a person's physical protection, entails coercion, or shows hatred or bigotry toward those protected groups, it is unlawful. And if the loss is not monetary, it is also a felony. Unintentional damage may arise when a dissatisfied employee creates a "harmless" virus that causes company disruption in any manner. Although it does not cause the same immediate financial harm as stealing confidential or financial records, it also has a financial impact due to wasted work hours and the resources spent to correct the issue.

Digital fraud, also known as cybercrime, is any crime involving a computer and a network. It's possible that the machine was used to commit a crime or that it was the intended target.

DOI: 10.1201/9781003280989-2

FIGURE 2.1   Introduction to cybercrime.

2.2   COMPUTER USER EXPECTATIONS

- Computer technology makes it faster and less costly to capture more data than was previously possible on paper. The increased utilization of electronic technology would allow for the capture of a greater variety of documents related to government functions. More purchases per person and per government employee would be allowed and encouraged by automated e-government facilities and programs, for example.
- Transactions and information are fine-grained. It would be possible to record—and preserve—all of the individual modifications made to a case file, as well as the final contents. Any person who "touches" (creates, shows, copies, or modifies) a record is tracked by content/document management systems. It's possible that the audit trail metadata would be very large. If data from government-operated sensors are automatically logged, new types of records will be generated.
- Time is money these days, and the quicker a computer user solves a problem, the easier they can go back to work. As a result, tech assistance desks are under a lot of pressure in answering users' requests as soon as possible. Help desk staff fail to verify users' identity or check whether they are allowed to conduct a requested action while under duress. In addition, even though computer users have been cautioned against it, they lend their login ID and password to colleagues who have lost theirs. This may incentivize businesses to seek access to data and information systems to which they are not otherwise permitted.

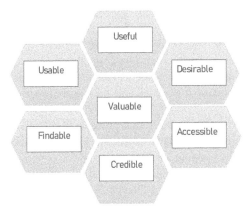

FIGURE 2.2   Factors contributing to a spike in computer-related vulnerability incidents.

Factors contributing to a spike in computer-related vulnerability incidents:

Vulnerability rises as complexity rises.

The computational world has grown in complexity. Hundreds of millions of lines of code link computers, networks, operating systems (OSs), software, Web pages, switches, routers, and gateways. Every day the difficulty of this world increases. As more devices are connected to a network, the number of potential entry points grows, raising the risk of security breaches. New risks are introduced by expanding and changing systems (Figure 2.2).

Business has progressed from a time where sensitive data are stored on a solitary mainframe computer in a locked room to a time when personal computers are connected to networks of millions of other computers, all of which are capable of sharing data. E-commerce, cloud networking, joint workgroups, multinational enterprise, and inter-organizational information networks have all exploded in popularity. IT has been pervasive, and it is also imperative that companies meet their objectives. However, keeping up with the speed of technical change, performing a successful ongoing evaluation of new security challenges, and implementing ways to deal with them is extremely challenging.

2.3   EXPANDING AND CHANGING SYSTEM

New risks are introduced by expanding and changing systems standalone machines, where sensitive data are kept on an isolated mainframe computer in a secured area, have developed in the corporate sector, to a world where personal computers and handheld devices link to networks of millions of other computers, both capable of exchanging knowledge. IT has become commonplace, and it is imperative that businesses meet their objectives. In contrast, IT organizations are finding it extremely difficult to keep up with the rapid speed of technical change, conduct continuous assessments of emerging threats, and develop strategies to address them.

BYOD (Bring Your Own Device) has become more popular nowadays. BYOD is a workplace policy that allows (and in some cases encourages) staff to access company computer services and software such as email, internal accounts, the corporate intranet, and the Internet through their devices (smartphones, tablets, or laptops). It is believed that providing employees access to familiar gadgets would increase their efficiency. However, since such devices are virtually definitely used for non-work activities (browsing the Internet, blogging, shopping, accessing social media sites, and so on), they are much more likely to be infected with malware than a device used exclusively for business. It's conceivable that ransomware will spread across the company.

Ransomware is a form of malware that prevents you from using your computer or viewing your files unless you pay a ransom or give the attacker images. It is possible to get corrupted by opening a malicious email file or visiting a hacked website. Viruses are pieces of programming code that allow a machine to act in an unintended and normally unpleasant way. They are typically masked as something else. About any virus is connected to a file, and the virus only executes when you open the infected file. It spreads to other computers when someone uploads an infected file or sends an email with a virus-laden attachment.

2.4 INCREASED RELIANCE ON COMMERCIAL SOFTWARE

An exploit is a computer attack on a system that takes advantage of a flaw in the system. Poor system design or implementations are frequently the cause of this attack. When a security flaw is found, software engineers work rapidly to produce and distribute a "fix," or patch, to remedy the problem. The patch, which can usually be downloaded from the Internet, is the responsibility of the system or application's users. Any time a patch is not installed, the user risks a security breach. The daily pace at which software vulnerabilities are found by companies across the globe is estimated to be between 7 and 382 per day. All of these bugs and potential vulnerabilities put developers who are in charge of security fixes under a lot of pressure. Keeping up with all of the required patches can be difficult. A zero-day intrusion occurs until the security industry or app developers are aware of the flaw or are ready to fix it. Although zero-day vulnerabilities can cause significant harm, few such attacks have been reported as of this writing.

2.5 SECURITY ATTACKS

Passive attacks and aggressive attacks are a helpful way of classifying security attacks, which is found in both X.800 and RFC 2828. A passive assault attempts to acquire or utilize information from the system without harming the system's resources. A successful attack tries to change the system's resources or disrupt its activity (Figure 2.3a).

Passive attacks

The opponent's objective is to intercept the information being exchanged. The publication of message content and traffic monitoring are two examples of passive attacks. The following diagram depicts the contents of a message being published. A phone conversation, an electronic mail address, or a transmitted file may all include relevant or private information. We don't want an enemy to know what's in these communications (Figure 2.3b).

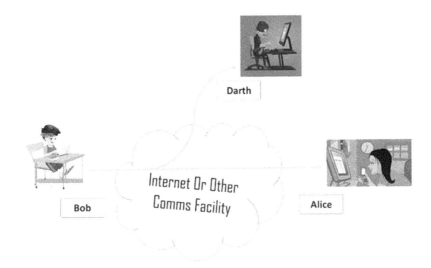

FIGURE 2.3A   Release of message contents.

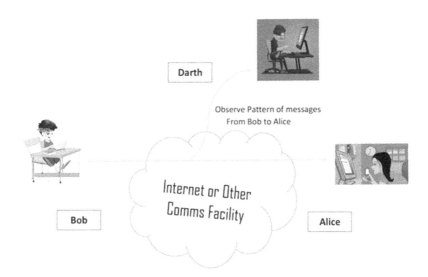

FIGURE 2.3B   Traffic analysis.

Traffic analysis is a more nuanced form of passive attack. Assume we had a means of masking the contents of communications or other intelligence flow so that adversaries could not retrieve the information from the communication even though it was intercepted. Encryption is a popular method for masking the content. And if we used cryptography, an adversary would still be able to view the sequence of these messages. The adversary may monitor the location and identity of transmitting hosts, as well as the frequency and duration of messages sent and received. This knowledge might help in figuring out what kind of contact was going on. Since passive attacks do not change the data, they are extremely difficult to detect. Letter traffic is always transmitted and retrieved in an unobtrusive manner, and neither the sender nor the recipient is aware that a third person has read the messages or detected the traffic pattern. However, by using cryptography, it is possible to avoid these attempts from succeeding. As a result, when coping with passive threats, the focus is on avoidance rather than identification.

Active attacks

Active attacks are grouped into four groups, each of which involves some tampering with the data stream or creating a fake stream:

Masquerade, replay, message alteration, and denial-of-service

When one person pretends to be another, it is called a masquerade. Each of the other successful attack types is normally used in a Masquerade attack. For example, once a legitimate authentication sequence has occurred, authentication sequences may be recorded and replayed, enabling a trusted entity with restricted permissions to acquire more privileges by impersonating a trusted entity with those privileges (Figure 2.3c).

The passive recording of a data device and eventual retransmission to achieve an unauthorized result is known as replay (Figure 2.3d).

To produce an unauthorized effect, some portion of a legitimate message is altered, or messages are delayed or reordered.

2.5.1 VIRUSES

The word "computer virus" has come to include a wide range of malicious software. A virus is a piece of computer code that enables a system to behave in an

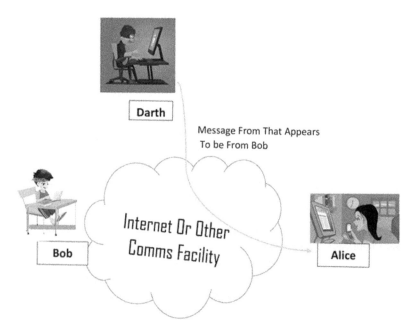

FIGURE 2.3C   Masquerade attack.

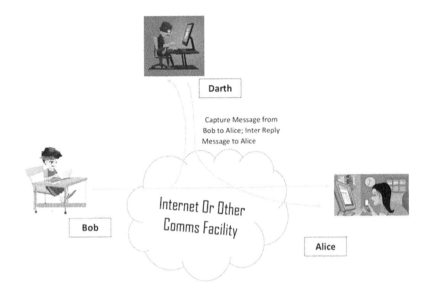

Darth

Capture Message from
Bob to Alice; Inter Reply
Message to Alice

Internet Or Other
Comms Facility

Bob

Alice

FIGURE 2.3D   Reply.

unexpected and usually unpleasant manner. It is typically masked as something else. Most viruses carry a "payload," or malicious content, which causes the device to behave abnormally. Real viruses do not propagate from one machine to another. When a computer user opens an infected email attachment, an infected program is installed on the machine or visits infected websites, and the infection spreads to other computers. Viruses propagate as a result of the actions of the "infected" machine user. Macro viruses have become a popular and simple type of virus to produce. Attackers corrupt documents and templates with a program written in an application macro language (such as Visual Basic or VBScript). Macros can change command functions or inject unnecessary words, numbers, or phrases into papers. A macro virus infects a user's application and then inserts itself in all subsequent documents created by the application. The virus deletes all data on the receiver's mapped discs if the email recipient clicks on the file.

Macros can change command functions or inject unnecessary words, numbers, or phrases into papers. After infecting a user's application, a macro virus will insert itself in all subsequent documents generated with the application.

Viruses migrate from one device to another, infecting them along the way. Viruses may have a wide variety of consequences, from slightly irritating to destroying data or applications to inducing denial-of-service (DoS) situations. About any virus is connected to an executable file, which ensures that although the virus may live on a system, it will not be active or transmitted until a user executes or opens the malicious host file or application. The viral code is executed at the same time as the host code. The host software, in most cases, continues to run after the virus has corrupted it. Such viruses, on the other hand, infect other applications with clones of themselves, effectively destroying the host software. Viruses propagate through the network, a disc, file sharing, or compromised email attachments as the device or document to which they are connected is passed from one device to another.

The Worst Computer Viruses of All Time

Klez is worth $19.8 billion.

ILOVEYOU is worth $15 billion.

WannaCry has a $4 billion price tag.

Zeus is worth $3 billion.

$2.4 billion for Code Red.

$1.2 billion for Slammer.

CryptoLocker is worth $665 million.

Sasser has a net worth of $500 million. Sven Jaschan, a 17-year-old German computer science major, created the Sasser worm.

2.5.2 WORMS

Computer worms, like viruses, replicate functioning copies of themselves and may do the same kind of harm. Unlike viruses, which spread by infecting a host file, worms are isolated programs that do not need a host program or human assistance to spread. To spread, worms either take advantage of a flaw on the target device or use social engineering to persuade users to run them. A worm infects a device by exploiting a system flaw and using the system's file-transport or information-transport features to fly unaided. To damage their victims, more sophisticated worms use cryptography, wipers, and ransomware technologies.

A worm assault on a company's machines will result in a lot of missed data and programs, as well as lost productivity from employees who are unable to utilize their computers, resulting in further loss in productivity while they attempt to restore data and programs, and a lot of effort from the workers to clear up the mess and get things back to normal as quickly as possible.

A worm is a harmful software that duplicates itself in a computer's active memory. Worms differ from viruses in that they can transmit without the aid of people, sometimes by emailing copies of themselves to other machines. The Code Red, SirCam, and Melissa worms were estimated to cost more than $1 billion each to patch, while the Conficker, Hurricane, and I LOVE YOU worms were expected to cost just over $5 billion each.

2.5.3 TROJAN HORSE

A Trojan horse, often known as a Trojan, is a malicious malware or software that seems to be normal but may take control of your computer. A Trojan horse is a computer software designed to damage, disrupt, steal, or otherwise harm your data or network.

A Trojan disguises itself as a genuine application or file in order to mislead you. It attempts to persuade you to install malware on your computer by tricking you into doing so. A Trojan will perform the purpose for which it was designed if it is deployed.

Although this is a misnomer, a Trojan is sometimes known as a Trojan horse virus or a Trojan virus. Viruses are capable of both execution and multiplication. It is impossible for a Trojan to do so. The customer is responsible for executing Trojans. Regardless, the words Trojan malware and Trojan virus are often interchanged.

It's important to understand how this infiltrator operates and what you can do to keep your computers secure, whether you call it Trojan malware or a Trojan virus.

As an example:

You might believe you've got an email from a friend and open what appears to be a valid attachment. You, on the other hand, have been duped. The email was sent by a cybercriminal, and the file you clicked on—and then downloaded and opened—went on to infect your laptop with malware. When you run the software, the malware will spread to other files on your machine and cause harm. How do you do it? It differs. Trojans are made to do a variety of tasks. However, you'll actually wish none of them were taking place on your computer.

After Trojans have been activated, hackers may use them to spy on you, steal your data, and get backdoor access to your device.

- Deleting data
- Blocking data
- Modifying data
- Copying data
- Interfering with the performance of computers or computer networks

Trojans, unlike computer viruses and worms, cannot reproduce themselves.

2.5.4 BOTNET

A botnet is a collection of computers linked to the Internet that has been infected with malware and may be controlled by hackers. Botnet assaults are carried out by cybercriminals who use botnets to carry out harmful activities such as password leaks, illegal access, data manipulation, and DDoS attacks.

Botnet administrators have the ability to command tens of thousands of computers at once, enabling them to conduct harmful operations. Cybercriminals get access to these computers by infecting them with specific Trojan viruses, and then installing command and control software to carry out large-scale destructive activities. These procedures may be automated to promote as many simultaneous assaults as feasible.

Botnet attacks can take a variety of forms, including:

- Distributed DoS (DDoS) attacks, which cause unplanned server downtime.
- Validating databases of stolen passwords that lead to account takeovers (credential-stuffing attacks).
- Giving an intruder access to a laptop and its network connectivity in order to intercept data from a Web application.

Under such scenarios, cybercriminals will offer links to the botnet network, also known as a "zombie" network, to other cybercriminals such that they may exploit it for their own sinister purposes, such as launching a spam operation.

Bots will record keystrokes, collect passwords, intercept and interpret packets, gather financial details, initiate DoS attacks, spam, relay, and open backdoors on the infected host, in addition to the worm-like capacity to self-propagate. Bots have all of the benefits of worms, but their infection vector is typically far more scalable, and they are often updated within hours of a new vulnerability being published. They've been known to use backdoors created by worms and viruses to gain access to networks with tight perimeter protection. Bots rarely make their existence known by scanning networks at high speeds, causing harm to network infrastructure; instead, they infect networks in ways that go unnoticed.

2.5.5 DDoS Attack

A DDoS attack is a concerted attempt to disrupt the normal operations of a targeted server or network by overwhelming the object or its neighboring networks with Internet traffic. DDoS attacks operate when they use a huge number of compromised machines as attack traffic sources. Exploited appliances have computers and other networked devices, such as Internet of Things (IoT) units. A distributed DoS attack is comparable to an unforeseen traffic congestion that clogs the roadway and stops normal traffic from reaching its destination. DDoS attacks are launched by connecting a group of computers to the Internet (Figure 2.4).

These networks are made up of malware-infected servers and other appliances (such as IoT devices) that can be manipulated remotely by an attacker. Individual robots are known as bots (or zombies), and a botnet is a collection of bots. The attacker may guide the assault by delivering remote instructions to each bot if a botnet has been created. When a botnet attacks a victim's server or network, each bot sends requests to the target's IP address, causing the server or network to become overloaded and resulting in a DoS attack. It may be difficult to differentiate attack traffic from regular traffic since each bot is a genuine Internet device.

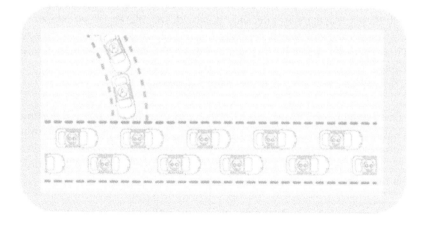

FIGURE 2.4   DDOS attacks are carried out with networks of Internet-connected machines.

2.5.6 ROOTKIT

A rootkit is a program or, more often, a collection of software tools that allows a threat actor to remotely access and manipulate a device or other machine. Although there have been legitimate applications for this form of software in the past, such as providing remote end-user service, most rootkits open a loophole on target networks, allowing malicious software such as viruses, ransomware, keylogger programs, and other forms of malware to be installed, or the device to be used for further network protection attacks. Rootkits also try to block endpoint antivirus software from detecting malicious software.

Rootkits may be placed on a target system through various methods, including phishing assaults or social engineering tactics that trick users into permitting the rootkit to be installed, which frequently gives remote hackers administrator access to the machine. Until it is activated, a rootkit gives the remote actor access to and control over almost any component of the OS. Unlike earlier antivirus programs that failed to detect rootkits, many antimalware applications now can look for and remove rootkits that are concealed within a device. Attackers may utilize the rootkit to execute programs, access logs, monitor user activity, and change the computer's settings. Rootkits are part of a blended attack, which often comprises a dropper, loader, and rootkit. The rootkit is installed using the dropper code, which can be triggered by clicking on a connection to a malicious Web site in an email or by opening an infected .pdf file. The loader software is launched by the dropper, which then deletes it. The rootkit is loaded into memory by the loader, and the machine is now infected. Rootkits are so cleverly built that it's impossible to even tell whether they're on a screen. The basic issue of attempting to identify a rootkit is that the OS that is actually in use cannot be relied upon to have accurate test results. Rootkits are classified into many categories based on how they infect, run, and persist on the target device.

A kernel-mode rootkit is designed to change the features of an OS. This kind of rootkit often injects its code and, in some instances, data structures into the kernel, the core of the OS. Many kernel-mode rootkits take advantage of the fact that OSs allow device drivers or loadable modules to run with the same level of system rights as the OS kernel, so rootkits are bundled as device drivers or modules to prevent detection by antivirus software. A user-mode rootkit, also known as an application rootkit, operates similarly to a regular user program. User-mode rootkits can be started up in the same way as most programs at device initialization, or they can be dropped into the system. The process is determined by the OS. A Windows rootkit, for example, usually focuses on modifying the basic functionality of Windows dynamic connection library files, but in a Unix environment, the rootkit can fully replace an entire program. The Master Boot Record of a hard drive or other storage device attached to the target machine is infected by a bootkit, also known as a boot loader rootkit. Bootkits have been successfully used to target applications that use complete disk encryption, so they can subvert the boot process and gain control of the machine after booting.

2.5.7 SPAM

Spam has been seen by someone who has spent more than a few seconds on the Internet. It seems to be an inextricable aspect of the Internet experience, something we take for granted. According to the answer, spam is still unrequested. It's inconvenient, usually promotional, sent to a huge number of people, and it comes whether or not you requested it. It's not spam if you signed up for marketing email and then grew tired of it. Spamming is the process of sending unsolicited mass communications, and a spammer is someone who engages in the practice. While spam is nearly usually commercial in nature, it isn't necessarily damaging or false (though it can be).

2.5.7.1   Spam's Different Types

Spam in the form of an email: standard spam. It clogs up your mailbox and diverts your attention away from the emails you want to post. Rest assured, it's all completely inconsequential. Search engine optimization (SEO) spam, also known as spam indexing, is the use of SEO techniques to boost the spammer's website's search results. We may categorize SEO spam into two types:

- Content spam: Spammers stuff their sites with common keywords that aren't relevant to their platform in order to boost their ranking in searches for certain keywords. Others will rewrite original material in order to make their pages seem more substantive and distinctive.
- Link spam: You've come across link spam if you've come across a blog entry or forum thread that's full of random connections. The spammer is attempting to push traffic to their website by using an SEO technique known as "backlinking."

Spam on social media: spammers can take advantage of the growing social nature of the internet, distributing their spam through bogus "throwaway" profiles on common social networking sites. Spam on mobile phones: it's a spam in the shape of an SMS message. Some spammers use push alerts in addition to spammy text messages to attract your attention to their deals.

Spam text messages: It's similar to email spam, except it's a lot faster. Spammers use instant messaging sites such as WhatsApp, Skype, and Snapchat to spread their messages.

2.5.8   PHISHING

Phishing is a kind of social engineering attack that involves obtaining sensitive information from individuals, such as login passwords and credit card details. When a hacker acts as a trustworthy person and convinces a victim to establish an account, send an instant message, or send a text message, this is known as social engineering. The customer is subsequently duped into clicking a harmful button, which may lead to the installation of malware, the freezing of a computer as part of a ransomware assault, or the disclosure of personal information. An effective attack has the potential to be catastrophic. Unauthorized transfers, money trafficking, and identity fraud all affect people. Phishing is often used to gain access to business or government networks as part of a broader assault, such as an Advanced Persistent Threat scenario. Throughout this scenario, employees are employed to penetrate defensive perimeters, spread malware within a restricted location, and get illegal access to sensitive information.

When a company is attacked, it usually suffers significant financial losses as well as a loss of market share, reputation, and customer confidence. Depending on the extent of the phishing effort, it may result in a security breach from which a company will be unable to recover.

2.5.8.1 Techniques Used in Phishing

- Phishing scams through email

    Phishing through email is a numbers game. Even if only a small percentage of recipients fall for the scam, an attacker who sends out thousands of fake messages will gain access to sensitive information and large amounts of money. As already mentioned, attackers use a variety of tactics to improve their success rates.

    For starters, they may take considerable efforts to make phishing communications that appear just like emails from a fake business. Because they utilize the same wording, fonts, labels, and signatures, the communications seem genuine.

    Additionally, attackers will try to persuade consumers to react by creating a feeling of urgency. As an example, as previously shown, an email may threaten account expiry and set a countdown for the recipient. The customer becomes less attentive and more susceptible to mistakes as the pressure is applied (Figure 2.5).

    Finally, inside-message connections mimic their legal counterparts but usually have a misspelt domain name or additional subdomains. The URL myuniversity.edu/renewal was updated to myuniversity.edurenewal.com in the example above. The similarity of the two addresses gives the appearance of a safe connection, leaving the receiver less aware of the threat.
- Phishing with a specific purpose

    In contrast to random app for consumers, it targets a single individual or business. It's a more advanced form of phishing that necessitates in-depth awareness of an organization's power system.

An attack could take any of the following forms:

1. A suspect researches the names of employees in a company's communications department and has access to the most recent project bills.
2. The offender sends an email to a departmental project manager (PM) with the subject line, updated invoice for Q3 campaigns, impersonating the marketing officer. The content, design, and logo are all similar to the standard email template used by the business.
3. A link in the email takes you to a password-protected internal document that is really a spoof of a stolen invoice.
4. The PM is asked to log in to see the text. The perpetrator obtains direct access to vulnerable parts of the organization's network by stealing his or her passwords.

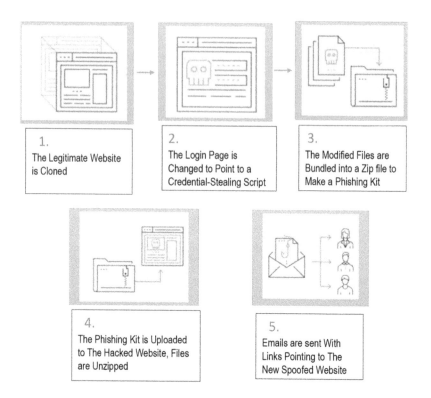

FIGURE 2.5   Life cycle of phishing kit.

2.6   PRIMARY PERPETRATORS

Hackers is a term used indiscriminately to describe perpetrators of cybercrime in general in a narrower context. As in other areas of violence, this raises the question of whether a distinction of criminal categories could also lead to distinct responses by the state in coping with those offenders.

The primary perpetrators of cyber fraud, as well as their goals:

2.6.1   HACKERS AND CRACKERS

2.6.1.1   Hackers

Hackers are a form of cybercriminals. Out of academic interest, hackers explore the limits of computer networks and see how they can obtain entry and how far they can go. They have a clear knowledge of IT and security features, and their inspiration derives in part from a need to learn more.

The word "hacker" has changed over time, and it now has a sinister connotation rather than the positive one it once had. Although a vocal group believes that hackers provide a benefit by finding security flaws, the majority of people now agree that hackers have no right to explore public or private networks.

Few hackers are smart and imaginative, but many are technically inept and are referred to be "lames" or "plot kiddies" by more experienced hackers. Hackers, surprisingly, have a wealth of opportunities at their disposal to refine their talents, including Internet chat rooms, Web forums, free hacking software, and even hacker conferences (such as DEFCON, an annual conference in Las Vegas). Valid licenses, such as CEH certificates, are carried by hackers. Hackers use programs or hacks to test a network's security and vulnerabilities. Hackers are often willing to share their information and never harm info. They work with a company to help secure their data and possess Internet security skills.

2.6.1.2 Crackers

A cracker is someone who maliciously hacks into or otherwise compromises the device functionality of remote computers. Crackers kill vital records, deny lawful user service, or create issues with their targets after gaining unauthorized access. Since their acts are malicious, crackers are quickly detected. Any cracker, whether professional or not, just knows a few tricks to steal records. These are the people who shield businesses from hackers. If they discover a loophole, they simply erase or harm the records. Crackers are corrupt and want to profit from illicit activities.

2.6.2 MALICIOUS INSIDERS

As seen in the opening vignette, a huge security risk for businesses is the disruptive insider, who is an ever-present and highly dangerous adversary. Companies are vulnerable to a variety of fraud threats, including misappropriation of money, asset stealing, fraud related to bidding procedures, invoice and payment fraud, electronic fraud, and credit card fraud. Internal fraud is normally the result of flaws in an organization's internal management procedures.

As a consequence, rather than being uncovered by monitoring processes, often frauds are uncovered by accident and by complaints from outsiders, settling payment disputes with vendors or manufacturers, or through a change of management. Frauds often require collusion, or coordination, between an individual and a third party.

Many statistics aim to estimate the danger posed by malicious insiders. Although the precise number can be debated, it is obvious that the danger exists; the only concern is the magnitude of the hazard. According to the CERT Insider Threat Centre3, a malicious insider threat to an organization is defined as a current or former employee, contractor, or other business partner who intentionally exceeded or misused authorized access to an organization's network, system, or data in a way that jeopardized the data's confidentiality, integrity, or availability. To be more accurate, this weakness applies to a malicious effort to compromise data or a data structure.

2.6.3 INDUSTRIAL SPIES

Industrial hackers steal trade secrets from their sponsor's rivals through illicit means. The Economic Espionage Act of 1996 protects trade secrets by making it a federal offense to use a trade secret for one's own or another's gain. Insiders, such as angry employers and ex-employees, are the most common thieves of trade secrets.

Competitive intelligence gathers material that is publicly accessible through legitimate methods. Financial accounts, trade publications, corporate filings, and printed interviews with company executives are used to compile and interpret information. Industrial espionage involves illegally collecting information that is not available to the general public. Participants might bug a conference space, wiretap the phones of key business executives, or steal sensitive test data from a research and development center. A dishonest corporation can recruit an industrial spy for a few thousand dollars to capture trade secrets worth tens of thousands of dollars. Since the employer's integrity (an intangible yet significant asset) will be seriously harmed if the espionage was exposed, the industrial spy prevents taking risks that would deceive his or her boss.

1. Industrial espionage is a relatively new phenomenon

    Theft of information for financial gain has been a fact for decades, but it exploded with the industrial revolution. All tried to snatch United Kingdom's secrets because it was the first to industrialize. As a result, both the export of heavy equipment and the emigration of professional workers is prohibited in the United Kingdom. The founding fathers of the United States were great supporters of stealing the secrets of United Kingdom. According to Alexander Hamilton and Benjamin Franklin, Americans should steal British inventions and immigrate professional labor. Samuel Slater, a well-known refugee, designed America's first water-powered textile mill using stolen British inventions, earning him the moniker "Slater the Traitor" in the English papers. Obviously, this burglary pattern persisted into the twentieth century. On a visit to a Ford plant in the United States in the 1920s, tourists from the Soviet Union stole blueprints and parts of a tractor. In the 1990s, a Gillette razor employee was found copying designs from a partner business. The robber was also charged with wire theft because he sent the trade secrets through fax and email. In a nutshell, corporate espionage has a long history. Globalization, global tourism, and the widespread use of the Internet are also recent concepts.

2. Hacking is not the only way for industrial espionage to take place

    For instance, espionage activities use all of the standard hacking and breaching techniques. Others, on the other hand, are hilariously low-tech. There are thousands of strategies that don't even include machines, such as dumpster diving, disrupting investor sessions, and making staff intoxicated at a pub.

3. Spy agencies are mostly involved in industrial espionage

    Private corporations are responsible for the various cyberattacks on industrial organizations, but universities and even workers are also involved. Employees who are unethical and know the monetary worth of the knowledge they have access to may attempt and sell it to the highest bidder, or they may be picked out and bribed.

4. Foreigners are mostly involved in industrial espionage

    It also occurs between corporations within the United States. It's not uncommon to see litigation filed by one Silicon Valley Corporation against another, alleging that an employee employed by the complainant took company secrets to their new workplace.

The rise of coworking spaces has created a new avenue for data theft. Rivals will easily set up shop in the same locations as any of the staff and then break into their networks through the Wi-Fi or physically breach computers after hours.

5. Industrial espionage is never appropriate

Interviewing an applicant for a position they don't want to give and manipulating the interview to learn business secrets are two examples. Unscrupulous corporations may even go undercover at business events in the hopes of obtaining intelligence from inebriated employees who believe they are conversing with a coworker.

6. Industrial espionage against the company is being carried out

True corporate espionage attempts are often directed at a third-party agency that has knowledge about the business. Your law or accounting firm, your partners, or even journalists covering your industry may be duped into divulging information about your company.

7. Industrial espionage hacking is the same as any other form of hacking

One feature that distinguishes industrial espionage attacks from, say, ransomware attacks is that corporate spies go to great lengths to ensure you never get to discover who they are. Most of the reasons they sometimes go unnoticed are because of this. Many businesses have been robbed and are unaware of it, so they are acutely aware of the risk. While your R&D lab is safe, what about the contract maker who makes your stuff, your patent attorney's office, and your accounting firm?

8. Industrial espionage is uncommon

Commercial espionage, according to the US Department of Justice, the Canadian government, NATO, and the United Nations, is on the rise. Between 2016 and 2018, more than half of all German businesses were victims of espionage, data stealing, or vandalism, according to studies undertaken by the German Association for IT, resulting in a $50 billion loss. According to the European Union, about 20% of all European businesses have been spied on (EU). Industrial espionage, it turns out, is fairly popular.

9. Theft of intellectual property is still used in industrial espionage

Theft of intellectual property is common in industrial espionage, particularly in industries where developing intellectual property is difficult or costly, such as aerospace or pharmaceuticals. Financial records, corporate or consumer details, or other confidential information is often the target.

2.6.3.1   How to Prevent Industrial Espionage

Following are a few precautions you can take to protect your company from industrial espionage attacks now that we've explained how risks can manifest:

- Identify the company's most sensitive details—the records that, if leaked, will spell disaster. Find out where it's kept and who has the keys to it, including third-party businesses and people.

- Establish a robust compliance policy and conduct security awareness training to secure sensitive properties. Pay attention not only to hacking but also to low-tech fraud techniques such as robbing computers or catching secrets on tape as a tourist. To put it another way, concentrate on both digital and physical protection.
- Use a cyber spectrum to simulate multiple threats to ensure technical protections and incident response plans are sound.
- Monitor file transfers, downloads, and email attachments with automated tracking tools to ensure staff aren't grabbing and deleting data they shouldn't.
- Collaborate with artificial intelligence to assess potential risks and react to current violations more efficiently.
- Ensure that workers use wiped phones and laptops with no data while traveling for work, particularly if they are going abroad. They can only use secure networks to access cloud-based data and a trusted virtual private network to access business data (VPN).

2.6.4   Cybercriminals

Cybercriminals are drawn to the use of information technologies because it allows them to target millions of possible victims, making it a modern and lucrative venue for them.

Financial advantage motivates cybercriminals to hack into corporate computers and loot; occasionally, money is transferred from one account to another, making it impossible for law enforcement agents to track them down.

Cybercriminals snatch and resell credit card numbers, personal names, and mobile phone IDs, among other types of cyber theft. They can afford to pay huge amounts of money to buy the technological skills and access they need from dishonest insiders because the opportunity for monetary gain is too great.

2.6.4.1   Cyber Terrorists and Hacktivists

Hacktivism is the use of malware to accomplish a political or social objective. It is a fusion of the terms hacking and activism. To advance such political or social agendas, a cyber attacker conducts computer-based attacks against other devices or networks in an effort to intimidate or coerce a nation. Although there is no direct distinction between cyber terrorists and hacktivists, cyber terrorists have more radical aims.

Cyber threats can quickly come from other countries thanks to the Internet, making identification and retribution even more difficult. Cyber terrorists employ tactics to kill or interrupt networks in order to inflict damage rather than collect intelligence. They are incredibly risky, treat them as though they are at war, have a high-risk tolerance, and pursue full effects. The following are key components of a multilayer protection vulnerability management process:

2.6.5   Cyber Terrorists

Hacktivism is a term that combines the terms hacking and advocacy to describe hacking for a political or social goal. A cyber attacker uses computer-based threats

Key Components of a Multilayer Security Vulnerability Management Approach:

Pillar	Microsoft's Efforts to Promote Trustworthy Computing
Security	Invest in the knowledge and technologies needed to provide a secure climate.
	To develop and implement safe computing, collaborate with law enforcement authorities, industry leaders, academia, and the private sector.
	Consumers should be educated on safe computing to build confidence.
Privacy	Invest in the information and technologies necessary to build a safe climate.
	To build and implement safe computing, collaborate with law enforcement, industry experts, academics, and the private sector.
	User education on safe computing will help build confidence.
Reliability	Build systems so that
	1. they continue to provide service in the face of internal or external disruptions;
	2. in the event of a disruption, they can be easily restored to a previously known state with no data loss;
	3. they provide accurate and timely service as needed;
	4. required changes and upgrades may not disrupt them;
	5. they have minimal software bugs when released; and
	6. They function as anticipated.
Business integrity	Be responsive
	• Accept blame for issues and take steps to resolve them.
	• Be transparent
	• Be frank with consumers, hold motivations transparent, follow up on commitments, and ensure sure customers realize where they are in relations with the business.

Difference between Industrial Spying and Competitive Intelligence:

S. No.	Industrial Spying	Competitive Intelligence
1	Commercial hackers procure trade secrets from rivals by illicit methods.	Competitive analysis is lawfully collected data gathered from publicly accessible databases.
2	Insiders, such as dissatisfied workers and ex-employees, are the most common thieves of trade secrets and intelligence.	Financial accounts, public filings, trade journals, and printed interviews with corporate executives are also used to collect information.
3	Spying on corporations is immoral and unlawful.	Competitive intellect is a socially and constitutionally acceptable market activity.

on other devices or networks to threaten or coerce a country into pursuing specific political or social goals. While there is no direct distinction between cyber terrorists and hacktivists, cyber terrorists have more radical aims. Cyber threats can quickly come from other countries thanks to the Internet, making identification and retribution even more difficult (Figure 2.6).

Cyber terrorists employ tactics to kill or interrupt networks in order to inflict damage rather than collect intelligence. They are incredibly risky, treat them as though they are at war, have a high-risk tolerance, and pursue full effects.

FIGURE 2.6   Cyber terrorists.

2.7   INDUSTRIAL SPIES AND COMPETITIVE INTELLIGENCE

Industrial manipulation is the illicit collection of intelligence in order to gain an advantage over a rival. For example, industrial theft occurs when a corporation learns that one of its rivals is going to deliver a product and obtains highly sensitive information from an internal source or by an illegal procedure. The explanation for this is that it does not have good intentions. Competitive Intelligence, on the other hand, is concerned with determining how successful the company is in comparison to the market. Recognizing the presence (and strengths) of your opponents benefits you in two ways:

1. Understand the business position.
2. Plan the next move to successfully compete with them.

To remain ahead of the market, any organization requires competitive intelligence in some manner.

But, before you start collecting competitive intelligence, you'll need a place to start, and competitive intelligence models are one way to do that.

Here are seven simple steps to gathering strategic intelligence:

1. Opposition
   You should be aware of the rivals. Not all items will have the same fea-
   tures as yours. Your competitor in this situation is the commodity that the
   target customer uses to meet their needs.
2. Objectives
   Find out why you're collecting data on other products/solutions. Are you
   attempting to:
   • Have a leg up on your rivals (in terms of revenues or outlets).
   • Expand the functionality of the offering.
   • Travel to new places.
   • Increase the appeal of your name.
   • Broaden your campaign reach.
   • Improve the quality of your customer service.
   • Expand the sales networks.
   • Lower the number of buyer rejections.
   In fact, there are a slew of other reasons to gather competitive intelli-
   gence data. To prevent doubt later, agree with partners on why you're doing
   this exercise ahead of time.
3. Areas of evaluation
   It's time to define observable metrics after you've worked out what you
   want to accomplish. If you want to increase sales enablement in your com-
   pany, for example, you can track the closing rate and sales cycle time. If you
   want to improve your marketing presence, keep an eye on your competitors'
   website traffic, pricing policy, and SEO strategy.
4. Participants
   Determine who in the company will absorb this strategic intelligence.
   Find out what stakeholders want from you. Your stakeholders are the
   Chief Technology Officer, VP of Software Engineering, and Head of User
   Experience while you're matching your products/solutions to those of your
   competitors. Stakeholders will differ depending on what you're trying to
   accomplish.
5. Sources of data
   Carefully choose the data sources. Hoover's and SEC filings are the best
   bets for business details such as sales, employee numbers, records, and
   annual results. Analyst reviews and expert interviews, on the other hand,
   should be your first options if you're searching for more classified info, such
   as consumers or future features. Depending on the source data you use, the
   strategic intelligence report can be strong or weak.
6. Templates for competitive intelligence
   It's a smart idea to keep plugging data into strategic intelligence models
   while you collect information so you don't lose track of it. These models
   must be created before the exercise and only exchanged with anyone that
   will be upgrading them.

7. Study publication

When the data collection is complete, share the findings with the stakeholders. Request input on any additional information that is needed, and remember to keep it updated regularly.

2.8   CYBERCRIMINALS

Cybercriminals are drawn to the use of information technologies because it allows them to target millions of possible victims, making it a modern and lucrative venue for them. Cybercriminals are motivated by financial gain to break into business servers to rob, often by converting funds from one account to another, leaving a hopelessly difficult path for law enforcement officers to trace. Cybercriminals snatch and resell credit card numbers, personal names, and mobile phone IDs, among other types of cyber theft. They can afford to pay huge amounts of money to buy the technological skills and access they need from dishonest insiders because the opportunity for monetary gain is too great.

Cyber terrorists and hacktivists

Hacktivism is the use of malware to accomplish a political or social objective. It is a fusion of the terms hacking and activism. To advance such political or social agendas, a cyber attacker conducts computer-based attacks against other devices or networks in an effort to intimidate or coerce a nation. Although there is no direct distinction between cyber terrorists and hacktivists, cyber terrorists have more radical aims.

Cyber threats can quickly come from other countries thanks to the Internet, making identification and retribution even more difficult. Cyber terrorists employ tactics to kill or interrupt networks in order to inflict damage rather than collect intelligence. They are incredibly risky, treat themselves as though they are at war, have a high-risk tolerance, and pursue full effects.

2.9   CYBER TERRORISTS

Hacktivism is the use of malware to accomplish a political or social objective. It is a fusion of the terms hacking and activism. To advance such political or social agendas and to threaten or compel a country, a cyber attacker uses computer-based assaults on other devices or networks. Although there is no direct distinction between cyber terrorists and hacktivists, cyber terrorists have more radical aims. Cyber threats can quickly come from other countries thanks to the Internet, making identification and retribution even more difficult. Cyber terrorists employ tactics to kill or interrupt networks in order to inflict damage rather than collect intelligence. They are incredibly risky, treat them as though they are at war, have a high-risk tolerance, and pursue full effects.

The media, the intelligence sector, and the IT industry have all been drawn to the challenge of cyber terrorism. Journalists, policymakers, and analysts from various disciplines have popularized a scenario in which advanced cyber terrorists use machines to hack into dams or air traffic control networks, wreaking havoc and jeopardizing not only millions of lives but also national security. For all of the doomsday scenarios, there has yet to be a single case of true cyber terrorism.

How serious is the danger posed by cyber terrorism? Because computers link the bulk of vital infrastructure in Western countries, cyber terrorism is a very serious danger. Hackers also show that individuals can obtain access to classified information and the activity of critical services, while not being inspired by the same agenda as terrorists. Terrorists might, at least in principle, follow the hackers' lead and, after breaking into government and private computer networks, cripple or at least disable advanced economies' military, banking, and service sectors. The increasing reliance of our communities on IT has developed a new type of weakness, allowing attackers to attack targets that would otherwise be impregnable, such as national security and air traffic control systems. The more sophisticated a country's technology is, the more susceptible it is to cyber assaults. As a consequence, the danger of cyber terrorism should be taken seriously. That isn't to argue that none of the worries voiced in newspapers, Congress, or other public forums are reasonable or practical. Some concerns are simply baseless, whereas others have become absurdly exaggerated. Furthermore, the distinction between the potential and actual damage inflicted by cyber terrorists has been overstated, and most hackers' relatively benign activities have been confused with the danger of pure cyber terrorism.

2.10   RISK ASSESSMENT

A risk assessment is a method for determining the security risks that an organization's systems and networks face from both internal and external threats. Such attacks will obstruct an organization's ability to achieve its core business goals. The aim of risk management is to determine which time and capital expenditures can help defend the company from its most possible and significant threats. An asset is any hardware, software, computer system, network, or database that is used by the enterprise to accomplish its business goals in light of an IT risk evaluation. A failure case is any event that has a detrimental effect on an object, such as a device being infected by a virus or a website being targeted by a DDoS attack. A general threat risk management mechanism is depicted in Figure 2.7.

The following are the steps in a general security risk evaluation process:

- Step 1: Identify the IT properties that the company is most worried about. Prioritization is usually assigned to properties that promote the organization's agenda and key business objectives.
- Step 2: Identify potential loss cases, risks, and threats, such as a DDoS attack or insider theft.
- Step 3: Estimate the number of incidents or the probability of each possible threat; certain risks, such as insider theft, are more likely to occur than others.
- Step 4: Determine the severity of each hazard. Is the danger likely to have a small effect on the enterprise, or does it prevent it from carrying out its task for an extended period of time?
- Step 5: Decide if each hazard can be mitigated so that it is either less likely to occur or has a smaller effect on the company if it does. Installing virus

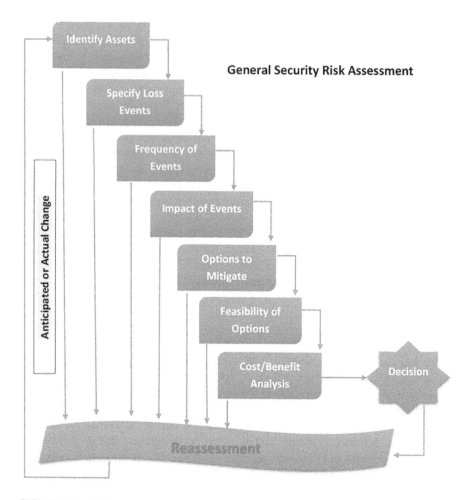

FIGURE 2.7   Risk assessment.

security on all machines, for example, makes it much less likely that a computer may get infected by a virus. Due to time and budget constraints, most
organizations tend to concentrate on risks with a high frequency (relative
to all other threats) and a high impact (relative to all other threats). To put
it another way, start with the risks that are most likely to exist and have the
greatest negative effect on the organization.
- Step 6: Evaluate the viability of putting the mitigation options into action.
- Step 7: Conduct a cost–benefit study and guarantee that the activities are
  profitable. Because no amount of money can ensure a flawless compliance
  scheme, businesses must weigh the possibility of a security breach against
  the cost of avoiding one. Fair assurance acknowledges that administrators
  must exercise their judgment to guarantee that the cost of control does not
  outweigh the system's gains or threats.

- Step 8: Determine whether to implement a particular countermeasure. If you decide not to employ a particular countermeasure, think whether the threat is really serious, and if so, choose a less costly option.

Computer Breaking Is Difficult due to Corporate Firewalls

1. A firewall acts as a barrier between an organization's internal network and the Internet, limiting network traffic in accordance with the company's access policies.
2. Firewalls may be created using software, hardware, or a mixture of the two.
3. All Internet traffic that aren't directly allowed onto the internal network are blocked.
4. Similarly, most firewalls may be set up to prevent internal network users from accessing such Web pages based on information such as sex or abuse.
5. Instant messaging, newsgroup access, and other Internet operations will also be blocked by most firewalls.
6. Installing a firewall will lead to complacency, which is a major security problem. A firewall, for example, would not prohibit a worm from infiltrating the network with an email attachment.
7. The majority of firewalls are set up to enable email and seemingly harmless attachments to enter their intended recipients.
8. Norton Personal Firewall, Comodo, and other top-rated firewall applications for personal computers are among the best.
9. A single-user license for firewall applications usually costs $30–$60.

A well-developed response plan should include the following key elements (Figure 2.8)

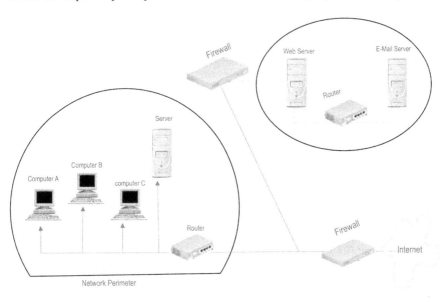

FIGURE 2.8   Firewall.

Notification of an incident

Determining whom to contact and who not to notify is an important part of any action strategy. The following are some of the questions to consider: Who needs to be informed inside the company, and what details does each individual require? What are the terms under which the corporation can approach its main clients and suppliers? How should the corporation notify them of a business interruption without alarming them unnecessarily? When can the FBI or local authorities be contacted? The majority of security professionals advise against disclosing specific details about a security breach in public media such as news stories, seminars, technical gatherings, and online chat groups. Without using systems linked to the corrupted infrastructure, all those working on the issue must be kept updated and up to date. The hacker may be watching these networks and emails to see what information is available about the security breach.

Evidence protection and activity logs

When it seeks to fix a security issue, an agency should log the aspects of the incident. Documentation collects important information for potential prosecutions and includes documents to aid in the eradication and follow-up stages of an incident. Keeping track of all machine activities, specific behavior (what, where, and who), and external conversations (what, where, and who) in a logbook is critical. Since this information can be used as testimony in court, an agency may develop a series of paper management protocols with the help of the legal department. It is frequently important to act immediately to contain an assault and prevent a terrible condition from worsening. The action strategy should specifically specify the procedure for determining whether an incident is dangerous enough to require the network to be shut down or vital systems disconnected. A successful action strategy considers how those decisions are taken, how quickly they are made, and who makes them.

Eradication of the disease

Before beginning the eradication effort, the IT protection team must gather and record all potential criminal information from the device, and then double-check that all required copies are present, accurate, and virus-free. It may be very helpful to create a forensic disk image of each compromised device on write-only media for later analysis and as documentation. The party must create a new backup until the virus has been eradicated. A list of all activities taken should be maintained in this period. This will be useful during the follow-up process, which will ensure that the condition does not occur again. Backing up sensitive software and data on a daily basis is important. Several organizations, on the other hand, have insufficient backup procedures in place and have discovered that they are unable to completely recover original data after a security event. Both backups should be made often enough to allow for a complete and easy restore of data if the original is destroyed. This procedure should be put to the test to ensure that it works.

Follow-Up on the Incident

1. Establish if the organization's protection was breached so that it would not happen again as part of the follow-up. Obtaining a technical upgrade from a device manufacturer is often the solution.

2. However, it's important to dig further than the quick fix to figure out why the incident happened. Why wasn't a basic security patch deployed before the incident happened if it could have avoided the incident?

3. After an incident, a study should be done to ascertain precisely what occurred and how the company reacted. One strategy is to write a written incident report that provides a comprehensive timeline of incidents as well as the incident's effects.

4. Any errors should be identified in this study so that they are not replicated in the future. The protection incident management strategy should be updated and revised based on the lessons learned from this incident.

5. The following are the essential components of a comprehensive incident report:
   - IP address and name of the concerned host computer(s).
   - The day and date of the incident's discovery.
   - How long did the event last?
   - The manner in which the occurrence was discovered.
   - How did you obtain access to the host computer?

A Brief Discussion about the Abused Vulnerabilities
- An assessment of whether the host was harmed as a result of the attack.
- The kind of information on the machine (customer, employee, etc.).
- Whether or not the information is intimate, private, or confidential.
- The length of time the machine was offline.
- The business's total influence.
- An estimation of the incident's gross monetary loss.
- A comprehensive timeline of the incidents surrounding the incident.

Creating a Security Policy Consists of the Following Steps:
1. A security strategy identifies an organization's security standards as well as the safeguards and sanctions used to satisfy them.

2. The roles and responsibilities of the organization's workers, as well as the activities that are expected of them, are outlined in a robust protection plan.

3. A compliance protocol specifies what must be done but not how it can be done. The specifics of how to achieve the policy's objectives are usually detailed in different manuals and protocol instructions.

4. The SANS (SysAdmin, Audit, Network, and Compliance) Institute's website has a range of security-related policy models that can help a company create appropriate security policies rapidly.

5. The following is a partial list of the SANS Institute's available templates:
   - Ethics Policy—this blueprint outlines the steps to establishing a corporate culture of transparency, trust, and dignity.
   - Information Sensitivity Policies—this sample policy outlines the criteria for classifying and protecting the organization's information according to its sensitivity level.

- Risk Assessment Policy—this template lays down the groundwork for the information management team to locate, evaluate, and mitigate threats to the organization's information systems that arise from doing business.
- Personal Communication Devices and Voice Mail Protocol Security Specifications—this example policy lays forth the security criteria for personal communication devices and voice mail.

6. Automatic machine rules can, as far as possible, represent an organization's written policy.

7. Using the setup options in a software application, automated system guidelines may also be implemented. For example, if a written policy mandates that passwords be updated every 30 days, all programs should be set up to automatically implement this policy.

8. Users, on the other hand, will always try to get around security policies or outright ignore them. Manufacturers of network routers, for example, advise customers to update the router's default password when they first set it up.

9. A hacker discovered a variety of routers around the world who are still using the default password and published a list of their IP addresses, enabling someone to obtain entry to the connected network and create havoc.

10. When introducing device protection controls, there are some trade-offs between ease of use and increased security; but, when a choice is made to promote ease of use, security incidents will increase.

11. Encryption techniques are becoming more accessible to end users as they get more sophisticated. Email attachments are a serious security concern that should be discussed in any company's security policies.

12. Even though a firewall and other security mechanisms are in place, sophisticated attackers can breach a network through email attachments. As a result, several businesses have decided to ban all incoming emails with a file attachment, significantly reducing their exposure.

Protective/Preventive Measures in Layers
Putting a corporate firewall in place

The most important security measure taken by companies is the installation of a corporate firewall. A firewall acts as a barrier between an organization's internal network and the Internet, limiting network traffic in accordance with the company's access policies. Firewalls may be created using software, hardware, or a mixture of the two. All Internet traffic that aren't directly allowed onto the internal network are blocked. Similarly, most firewalls may be set up to prevent internal network users from accessing such Web pages based on information such as sex or abuse. Instant messaging, newsgroup access, and other Internet operations will also be blocked by most firewalls. Installing a firewall will lead to complacency, which is a major security problem. A firewall, for example, would not prevent a worm from infiltrating the network with an email attachment. The majority of firewalls are set up to enable email and seemingly harmless attachments to access their intended recipients.

Intrusion detection systems

An intrusion detection system (IDS) is a software or hardware that monitors system and network resources and events, and notifies network security professionals when it detects network activity that attempts to circumvent security measures in a networked computer environment. Such behaviors typically indicate an effort to compromise the system's security or restrict network resource capacity. Intrusion detection may be divided into two categories: knowledge-based methods and behavior-based approaches.

Knowledge-based intrusion detection programs keep track of individual threats and device flaws, such as repetitive unsuccessful authentication attempts or recurring attempts to update a program to a computer. An alert is activated if such an attempt is observed. A behavior-based intrusion detection framework models typical behavior of a system and its users using reference data acquired through various ways. If a deviation is discovered, the IDS compares real behavior to the model and issues an alarm. Two instances include unusual traffic at odd hours or a customer in the Human Resources Department using accounting software he or she has never seen before.

Antivirus software installation on personal computers

Antivirus software should be installed on each user's device to search the memory and disk drives for viruses on a daily basis. An antivirus program looks for a virus signature, which is a fixed sequence of bytes that signals the existence of a specific virus. If the antivirus software detects a virus, it notifies the user and can disinfect, erase, or quarantine any files, folders, or disks that have been infected by the malicious code. Good antivirus software scans critical device files as the computer boots up, watches the computer for virus-like behavior on a continuous basis, scans drives, scans memory when a program is run, scans programs when they are downloaded, and scans email attachments until they are opened. Norton Antivirus from Symantec and McAfee Personal Firewall are two of the most commonly deployed antivirus security tools.

Putting in place protections against malicious insider attacks

User accounts that are still operating after workers have left a business pose a security danger. To minimize the risk of a hostile insider attack, IT workers can erase leaving employees' and contractors' computer profiles, login IDs, and passwords as soon as possible. Organizations must also carefully identify job duties and adequately distinguish core tasks such that no one individual is responsible for completing a mission of high security consequences. Allowing an individual to both execute and accept buying requests, for example, will be counterproductive. An employee could enter huge invoices on behalf of a "friendly vendor," accept the invoices for payment, and then leave the organization to share the money with the vendor. In addition to splitting responsibilities, many companies rotate staff in critical roles regularly to avoid insider crimes. Another critical precaution is to create tasks and user identities so that users have only the authority they need to carry out their duties.

Defending from cyber-terrorist attacks

In the face of rising cyber terrorism threats, businesses must be mindful of the tools available to assist them in combating this serious danger. The Department of Homeland Security (DHS) is in charge of "securing civilian government data networks and collaborating with business, state, provincial, tribal, and territorial governments to protect sensitive resources and information systems" for the federal government. The DHS aims at "analyzing and mitigating cyber risks and vulnerabilities; delivering vulnerability warnings; and organizing the response to cyber events to ensure that our devices, networks, and cyber infrastructure stay safe," according to the department's website. Defending against the Most Severe Internet Security ThreatsWell-known bugs are used in the vast majority of effective malware attacks. Since computer criminals know that certain organizations take a long time to address issues, searching the Internet for compromised applications is a viable attack technique. US-CERT publishes a list of the most common, high-impact vulnerabilities that are submitted to them on a daily basis. This description is available at www.us-cert.gov/current. To fix these problems, you'll need to apply a proven security patch to keep the programs and operating systems up to date. Those in charge of information protection must prevent attacks based on these flaws a top priority.

Auditing IT security on a regular basis

A compliance assessment to see that a company has a well-thought-out security strategy in place and is following it. For example, if a policy requires all users to update their passwords every 30 days, the audit must examine how well the policy is followed. The audit should also look at who has access to specific applications and records, as well as the extent of authority that each individual has. It's not uncommon for an investigation to find that so many people have access to sensitive information and too many people have skills beyond what they need to do their work. A successful audit will provide a list of issues that need to be corrected in order to ensure compliance with the security policies.

Computer Forensics or digital forensics is the study of how computers can be used to solve crimes. Digital forensics is a discipline that incorporates aspects of law and computer science to classify, obtain, analyze, and protect data from computer systems, networks, and storage devices in such a way that the data are preserved in their integrity and admissible as evidence in a court of law. In relation to a criminal case or civil action, a data forensics investigation can be launched. It can also be used to retrace measures taken when data are missing, to quantify harm after a computer incident, to examine the unlawful leakage of personal or organizational sensitive data, or to validate or determine the effects of industrial espionage, among other things. Successfully combating cyber fraud in a court of law requires proper treatment of a computer forensics case. A data forensics investigator's status in a court of law is also enhanced by rigorous experience and qualifications. Electronic forensics certifications include the CISSP (Certified Information Systems Security Professional), CCE (Certified Computer

Examiner), CSFA (CyberSecurity Forensic Analyst), and GCFA (Global Computer Forensic Analyst; Global Information Assurance Certification Certified Forensics Analyst). Experts who have learned computer investigative techniques as well as the use of Guidance Software's incase computer forensics tools are certified as EnCE Certified Examiners. Computer forensics degrees are available from a variety of institutions (both online and traditional). Accounting, especially auditing, preparation should be included in those degree programs, as it is very helpful in fraud investigations. A digital forensics expert must be familiar with the different rules that govern the collection of criminal evidence.

2.11 TRUSTWORTHY COMPUTING

Trustworthy computing is a computing approach that provides stable, confidential, and consistent computing interactions based on sound market practices; this is what today's companies require. Everybody in the computer industry (software and hardware designers, contractors, programmers) understands that this is a top priority for their clients. Microsoft, for example, has committed to a trustworthy programming campaign to increase confidence in its tech offerings, as shown in Figure 2.9.

Every device or network's security is a blend of technology, regulation, and individuals, and it necessitates a wide variety of activities to be successful.

> Ultimately, society wants computer systems to be trustworthy—that is, that they do what is necessary and demanded of them through disruptions in the climate, human user and operator mistakes, and aggressive party assaults, and that they do not do something else.

A good protection policy starts with an assessment of threats to the servers and network of the company, the identification of measures to fix the most critical vulnerabilities, and the education of end users about the dangers involved and the steps they must take to avoid a security incident. Through enforcing security policies and protocols, as well as efficiently employing accessible hardware and software resources, the IT security community must lead the initiative to deter security breaches. However,

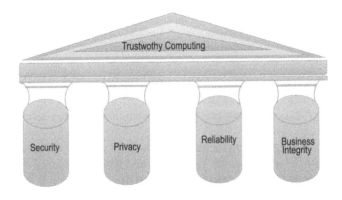

FIGURE 2.9 Trustworthy computing.

since no detection mechanism is flawless, processes and protocols must be constantly checked in order to detect a potential intruder. In the event of an attack, a specific response plan must be in place that covers warning, evidence preservation, operation record maintenance, containment, eradication, and recovery.

A protection policy is a text that states in writing how an organization wants to secure its physical and IT properties in the workplace. A security policy is sometimes referred to as a "living paper," meaning that it is never completed and is changed as technologies and employee needs alter. A reasonable use policy, an overview of how the organization aims to teach its staff on securing the company's properties, an outline of how security measurements can be carried out and implemented, and a process for reviewing the security policy's efficacy to ensure that appropriate changes are made are all part of a company's security policy.

The rules and procedures for all people accessing and using an organization's IT assets and resources are outlined in an IT Security Policy. Effective IT Security Policy is a representation of the company's culture, encompassing rules and procedures based on workers' views of their data and job. As a result, an effective IT security plan is a unique document for each organization, crafted from citizens' perspectives on risk tolerance, how they see and trust their data, and the information's resulting availability. As a consequence, many companies will discover that a conventional IT security strategy is inadequate since it ignores how workers utilize and share information both inside the company and with the general public.

The secrecy, transparency, and usability of systems and records used by an organization's members are the goals of an IT security strategy. The CIA triad is made up of three principles:

(a) Availability is a state of the environment in which designated users have constant access to said assets. (b) Confidentiality guarantees that assets are protected from unauthorized entities. (c) Integrity ensures that asset alteration is done in a defined and authorized manner.

The IT Security Policy is a dynamic document that is updated regularly to reflect evolving business and IT requirements. Organizations such as the International Organization for Standardization (ISO) and the National Institute of Standards and Technology (NIST) in the United States have produced standards and best practices for the development of military strategy. According to the National Research Council (NRC), every business agenda could discuss the following:

1. Objectives
2. The scope of the project
3. Specific Goals
4. Regulatory responsibilities and steps to take in the case of noncompliance

Any IT security strategy should include sections devoted to ensuring that the organization's industry's regulations are followed. The PCI Data Security Standard and the Basel Accords, as well as the Dodd-Frank Wall Street Reform Act, the Consumer Protection Act, the Health Insurance Portability and Accountability Act, and the Financial Industry Regulatory Authority, are examples of legislation in the United

States. Many of these governmental entities have their own IT security protocols that are freely accessible.

A company's protection strategy will affect its actions and path, but it will not alter its plan or purpose. As a result, it's critical to write a policy that's based on the organization's current cultural and institutional structure to ensure that successful efficiency and innovation continue, rather than a generic regulation that prevents the company and its citizens from achieving their purpose and goals.

The significance of executive preparation and education:

The significance of workforce preparation and education cannot be overstated. Employee capability development, talent retention, and career satisfaction are all aided by training and education. Employees gain the skills they need to do their work more efficiently through training and education. As a result, they are more likely to achieve their full potential. Employees who have been well educated are more effective and make fewer errors.

Since it affects efficiency, sales, and corporate productivity, training and teaching workers may help a company grow. Unfortunately, many businesses slash their training expenses first as sales fall or the economy slows. Do not fall into this pit because you can end up losing the company's long-term value. The below are several scenarios in which your employees' education and preparation are critical.

Orientation: Training your new hires would ensure that they are aware of their responsibilities as soon as they begin working for you. Employees are given knowledge about workplace practices, strategies, and corporate theory during orientation. This is an important step towards developing a positive working relationship with them. You'll save money and resources by not having to answer inquiries about routine job practices or rules if you train new hires.

Job skills preparation: Both staff and managers benefit from job skills training. It aids in the development of employees' talents, thus improving morale and efficiency. Employees with higher working satisfaction and productivity are more likely to do well in their careers. Offering work skills preparation to your workforce can help you get a better return on investment as an employer.

Workplace protection: As an employer, you should have a healthy working space for your workers. As a result, occupational safety training is important. Since workplace crime is on the rise, you can teach your workers how to handle potentially dangerous conditions. You can not only provide safety training but also do routine inspections to demonstrate the value of safety training.

Professional growth: You can find workers who can carry on more roles in your company by providing them with training and education. Employees who wish to advance their careers through professional advancement and promotion opportunities will benefit from training. Since succession planning is an integral part of long-term planning, preparing and training your staff should be a priority in your company's future objectives. Professional learning is a fantastic way to keep your workers motivated at every point of their careers. Investing in the company's future leaders demonstrates that you are committed to your company's mission and values.

Maintaining the necessary training and education would help the company stay competitive. By creating a robust training schedule, you can reduce and monitor the costs of training your staff. You can save much more money by using online employee training tools. Any online tools will help you train your employers in addition to helping you train your employees.

Identifying and addressing employee weaknesses:

The majority of employees have certain weaknesses in the job. You may build the skills that each employee needs with the help of a training program. A learning program improves the quality of all employees by ensuring that they have the same level of knowledge and experience. This attempts to remove any weak spots in the company when basic job duties are delegated to others.

Workers would be able to take over for one another if needed, operate in teams or work alone without the need for constant support and monitoring if the necessary preparedness was provided.

Employee performance has improved:

Employees that have received the necessary training are well prepared to perform their duties. She improves her knowledge of basic work safety and practices. The employee's morale may even improve as a result of the training since she has a greater knowledge of the company and the requirements of her job. This self-assurance will encourage her to work harder and come up with fresh ideas to help her achieve.

Your workforce will stay on the cutting edge of industry trends thanks to ongoing preparation. Competent employees who keep up with evolving market expectations help the company maintain its status as a forerunner in the region and a strong competitor.

Structured education and training

Employees with a consistent history and background expertise benefit from a formal training and learning schedule. The importance of continuity is especially important when it comes to the company's basic policies and procedures.

Both employees must adhere to the company's policies and procedures. This includes defense, prejudice, and administration, among other things. All personnel are provided daily training in these areas, ensuring that they are at least exposed to the content.

Job satisfaction among employees:

Employees who have access to training and growth services have an edge over others who are left to sort out training options on their own at other businesses. A company's commitment to provide training tells its workforce that they are respected. The teaching fosters a positive work environment.

Employees can seek access to training they would not be aware of or find out on their own. Employees who are valued and pushed by educational opportunities are more likely to be satisfied with their jobs.

REVIEW QUESTIONS

1. What action must be taken in response to security policy?
2. Discuss why computer-related security incidents have rapidly increased in recent years.
3. Explain how vulnerabilities can be reduced.
4. What are some of the most frequent computer security threats?
5. Explain why computer incidents are so prevalent.

6. With the help of a neat diagram explain risk assessment in detail.
7. "Educating employees is always beneficial for an organization" - Are you agree with the statement? Justify your answer.
8. Explain Prevention. What are the protective measures for prevention?

MULTIPLE CHOICE QUESTIONS (MCQs)

1. **Which of the following does not qualify as a kind of cybercrime?**
   a. Installing antivirus for protection
   b. Damage to data and systems
   c. Forgery
   d. Data theft
2. **Which of the following is not an example of cyber-crime using a computer as a weapon?**
   a. Spying someone using keylogger
   b. Credit card fraudulent
   c. Pornography
   d. IPR Violation
3. **What is the name of the IT legislation passed by India's legislature?**
   a. India's Technology (IT) Act, 2000
   b. India's Digital Information Technology (DIT) Act, 2000
   c. India's Information Technology (IT) Act, 2000
   d. The Technology Act, 2008
4. **In India, what is the punishment for stealing computer data, money, or software source code from any business, person, or another source?**
   a. One year of imprisonment and a fine of Rs. 100,000
   b. Three years of imprisonment and a fine of Rs. 500,000
   c. Six months of imprisonment and a fine of Rs. 50,000
   d. Two years of imprisonment and a fine of Rs. 250,000
5. **What types of cybercrime does section 66 of the Indian IT Act include, as well as the rules and penalties that apply to them?**
   a. Stealing hardware components
   b. Stealing data
   c. Putting antivirus into the victim
   d. Cracking or illegally hacking into any system
6. **A computer _____ is a malicious software that copies itself to another program in order to reproduce itself.**
   a. program
   b. application
   c. virus
   d. worm
7. **The virus hides itself from getting detected by _____ different ways.**
   a. 2
   b. 3
   c. 4
   d. 5

8. _____ is difficult and complicated to delete this virus since it infects the master boot record.
   a. Boot Sector Virus
   b. Polymorphic
   c. Multipartite
   d. Trojans

9. _____ installs itself and lies locked in the computer's memory. It remains centered on the various types of files it infects.
   a. Boot Sector Virus
   b. Direct Action Virus
   c. Polymorphic Virus
   d. Multipartite Virus

10. Any of the following theories may not meet the criteria for why people develop computer viruses?
    a. Research purpose
    b. Pranks
    c. Identity theft
    d. Protection

ANSWERS TO MCQs

Q1: a, Q2: b, Q3: c, Q4: b, Q5: d, Q6: c, Q7: b, Q8: a, Q9: b, Q10: d

3 Organization Privacy

LEARNING OBJECTIVES

- To be familiar with the basic ideas of privacy in different foreign jurisdictions of cyber laws.
- Understand organization's obligations when sharing data internally, internationally, and across borders.
- Understand the rights that individuals have with regard to their personal data.
- Understand cyber security laws around the world.

3.1 INTRODUCTION TO CYBER SECURITY

Over the last half-century, the information and communications technology (ICT) market has changed dramatically. Technology is pervasive and is more and more integrated into nearly every aspect of contemporary culture. ICT systems and materials are inherently interdependent, and a failure of one may have far-reaching consequences for others. Experts and policymakers have voiced growing concern in recent years about defending ICT networks from cyber attacks, which many experts predict will escalate in magnitude and intensity in the coming years.

Cyber protection refers to the act of safeguarding ICT networks and their contents. Cyber protection is a vast and arguably nebulous idea that can be useful but defies clear meaning. It has also been mixed up with other terms such as anonymity, knowledge sharing, intelligence collection, and monitoring in the past. Cyber protection, on the other hand, can be a valuable weapon for defending privacy and avoiding unwanted surveillance, and knowledge exchange and intelligence collection can help with cyber security. Benefit assessment for information infrastructure is regarded as critical to successful Cyber defense. Any attack's risks are determined by three factors: challenges (who is attacking), flaws (how they are attacking), and consequences (what the attack does). Many cyber attacks are minor, but a successful assault on critical infrastructure (CI)—the majority of which is owned by the private sector—could have major ramifications for public security, economy, and individual citizen's livelihood and protection. Typically, reducing those threats entails eliminating hazard vectors, fixing vulnerabilities, and reducing their consequences.

In terms of cyber defense, the federal government is responsible for both defending federal infrastructure and aiding in the protection of nonfederal systems. Both government departments have data security responsibility over their own networks under existing legislation, and many have sector-specific CI responsibilities. Federal departments spend more than 10% of their annual ICT budgets on cyber security

DOI: 10.1201/9781003280989-3

on average. More than 50 statutes deal with different facets of data defence, and additional legislation has been discussed since the 111th Congress. Executive Order 13636 and Presidential Policy Directive 21, both issued in February 2013, discuss CI cyber protection through voluntary public–private coordination and the application of established regulatory authorities. Four bills passed in December 2014 cover federal ICT security, the Department of Homeland Security's (DHS) cyber security staff, and DHS information-sharing operations. Other legislations fix concerns such as intelligence sharing more generally, research and advancement, CI security, data loss reporting, and cybercrime rules, among others.

The executive orders and pending regulations were mostly intended to resolve a number of well-known near-term cyber protection needs. Such requirements, however, occur in light of more complex long-term issues such as architecture, compensation, compromise, and the environment. The 114th Congress's legislation and executive orders will have a big effect on these issues. Cyber security refers to the methods, techniques, and processes used to secure the confidentiality, integrity, and functioning of information systems, networks, and data against cyber attacks or unauthorized access. Cyber defense's main aim is to protect all business assets against both external and internal assaults, as well as natural disaster-related disruptions.

Since an organization's facilities are made up of a variety of diverse programs, a strong data defense policy necessitates concerted activities through all of its systems. As a result, cyber defense has the following subdomains:

- Security of the application

    The implementation of different protections within all technologies and facilities used within an enterprise against a broad variety of threats is known as application protection. To reduce the risk of any unwanted entry or alteration to application infrastructure, it necessitates developing protected application architectures, writing secure code, applying strong data input authentication, hazard modeling, and so on.
- Data protection and identity management

    Identity protection refers to the mechanisms, procedures, and practices that allow legitimate persons to access information systems within an enterprise. Implementing strong information management systems to guarantee data integrity at rest and in transit is part of data protection.
- Network safety

    The implementation of both hardware and software frameworks to defend the network and facilities from unwanted access, delays, and misuse is known as network protection. Effective network management aids in the protection of an organization's infrastructure from a variety of external and internal risks.
- Mobile safety

    Mobile protection refers to safeguarding all corporate and personal data stored on mobile devices such as cell phones, computers, tablets, and other similar devices from risks such as unauthorized entry, computer failure or theft, ransomware, and so on.

- Cloud safety

  Cloud protection refers to the development of stable cloud architectures and software for businesses that use AWS, Google, Azure, Rack space, and other cloud service providers. Defense from different risks is ensured by effective design and ecosystem configuration.
- Corporate management and disaster recovery plans (DR&BC)

  DR&BC works with procedures, reporting, warnings, and preparations to help businesses prepare for maintaining business-sensitive services online after and during a crisis, as well as resuming some activities and systems.
- Information for users

  Formal instruction on information security issues is critical for increasing knowledge of industry best practices, corporate processes, and regulations, as well as tracking and reporting malicious activity.

3.2 GLOBAL CYBER SECURITY AND RELATED LAWS IN THE UNITED STATES

Cyber protection is mandated by a number of federal and state regulations. The Federal Trade Commission (FTC) has been especially involved in this region, using its regulatory power under Section 5(a) of the FTC Act, which applies to unethical and misleading acts, to compel businesses to enact compliance measures. Since 2002, the FTC has filed more than 80 compliance cases against businesses that it claims have neglected to take appropriate security precautions.

The Cyber Security Information Sharing Act (CISA) has two major consequences. For starters, it enables businesses to track network activity and take protective measures on their own networks. Second, it promotes the exchange of cyber-threat data between businesses and the government.

FIGURE 3.1 Global cyber security and related laws in United States.

Any federal regulations, on the other hand, are industry-specific. The Gramm-Leach-Bliley Act (GLBA) and its implementing legislation, for example, mandate financial institutions to enforce established rules and processes that are adequately designed to guarantee that customer records are protected and kept private, as well as protect against risks and unwanted access and use. Encrypted health details in the hands of such "insured agencies" and their "financial partners" are subject to electronic protection provisions under the Health Insurance Portability and Accountability Act (HIPAA).

Several states have enacted legislation implementing protection requirements. The majority of these laws demand some kind of "fair protection." Companies that possess or lease personal information must comply with Massachusetts laws, which include implementing a written authentication policy and encrypting data in transit through public networks and on all handheld devices. New York also passed the SHIELD Act, which establishes a requirement for fair data protection and specifies various steps that should be used to meet the standard. If claimants prove that the affected company failed to enforce and uphold fair protection policies and standards, to safeguard personal information that is relevant to the nature of the information, the California Consumer Privacy Act (CCPA) provides a data breach right of redress for California customers, with statutory fines ranging from $100 to $750 per consumer and per incident.

CISA, a portion of the DHS and the federal agency in charge of securing vital infrastructure in the United States, was established by the Cyber Security and Infrastructure Security Service Act. To secure vital facilities, CISA coordinates efforts between government and private sector entities. The federal government has provided sector-specific guidelines for essential infrastructure operators, and there are comprehensive legislative and regulatory standards in the nuclear, chemical, electrical, government procurement, aviation, and other industries.

Some US rules are much more prescriptive than others, in addition to general fair protection standards. At the state level, Massachusetts' electronic security laws and the New York SHIELD Act, for example, have comprehensive information security provisions, and the New York Department of Financial Services (which oversees institutions including banks and insurance companies) has additional standards.

The notification of incidents is required in all 50 states and 4 territories, with the majority of these laws requiring reporting to state regulators. The type and extent of information that must be disclosed differs by state or jurisdiction. For example, in Massachusetts, organizations reporting a breach to state regulators must provide information about (a) the nature of the security breach or unauthorized acquisition or use, (b) the number of people in Massachusetts who were impacted by the incident, (c) whatever actions done in response to the incident, (d) the name of the company that reported and was affected by the incident, and (e) the date the breach occurred.

Cyber protection rules did not have much weight in the previous century. At the moment, the form of cybercrime being perpetrated was not as serious as it is now. The rules were similar to copyright rights or software theft laws at the time.

However, the challenge has grown, and even more serious cybercrime has become the norm. These offences vary from the use of ransomware to outright treason.

To combat and prevent such offences, serious action must now be taken. Increased regulatory intervention has resulted from the increased challenge.

Prior to 2015, the US federal government was unaware of a number of attempted data hacks at private institutions. Through the passage of the Cyber Security Act of 2015, much of that shifted. Congress finally passed law allowing businesses in the United States to exchange personal information relating to computer protection with the government after several attempts. This database may be used by the government to punish criminals.

It is unlawful to perform the following activities everywhere around the globe under cyber protection rules:

- Breaking into computer systems, accessing illegal data, altering or deleting the data
- Stealing private information
- Unauthorized publishing or use of communications
- Criminal violation of copyright
- Spreading false information
- Sexual exploitation of minors

Numerous other offences committed over the Internet have been criminalized under different categories of the legislation.

Cyber security is addressed in the United States by industry-specific initiatives, general oversight, and private sector participation. Cyber security standards are implemented in many ways at the state or federal level.

The FTC is the official federal body in charge of implementing the ban on "unfair and misleading acts or practices." Using this jurisdiction, the FTC also imposes minimum security standards on organizations that gather, retain, or store personal information about consumers.

The FTC released "Start with Protection" guidelines in June 2015. The lessons learnt by the FTC from over 50 data protection compliance measures brought by the FTC since 2001 were correctly described in this guidance. Companies can implement a set of ten lessons learnt, ranging from authentication controls to network segmentation, according to this guideline. An FTC order imposed against a corporation for supposedly "unreasonable" protection measures in excess of the FTC Act was vacated by a federal appeals court in mid-2018.

The court ruled that the FTC's injunction did not require the corporation to stop engaging in any particular unjust actions or activities. Rather, it only required it to provide a "comprehensive information protection scheme" that was "reasonably structured to safeguard the security, confidentiality, and integrity of personal information obtained by or about customers."

The court sidestepped the larger question of whether the suspected security flaws amounted to "unfair" market practices under the FTC Act. Sections of the FTC's previous data protection consent orders were called into doubt by the ruling. The FTC's approach to future data protection compliance activities could change as a result of this.

- HIPAA stands for Health Insurance Portability and Accountability Act (1996)

  HIPAA was signed into law by President Bill Clinton in 1996.

  There was no uniform procedure for safeguarding confidential personal information that was retained by healthcare organizations prior to HIPAA. There were no established security best practices. Most of the reasons for the lack of data protection practices in the healthcare sector were that patient records were traditionally kept on paper.

  The healthcare sector was scrambling to step away from paper notes to become more productive just before HIPAA was enacted. The desire to obtain and move patient records soon arose from the need to become more effective.

  Since there was a pressing need to move to electronic medical records, a slew of businesses sprung up to meet the demand. Many of these businesses saw security as an afterthought. The government soon recognized the need for laws to implement security requirements.

  HIPAA's main goals are to:
  - Modernize how healthcare information is handled and processed,
  - Ascertain that hospitals, insurance companies, and other health-related organizations keep sensitive personal information safe,
  - Resolve the issue with healthcare insurance limits.
- The Gramm–Leach–Bliley Act (GLBA) was enacted in 1999

  In 1999, the GLBA was signed into law. The Financial Services Modernization Act of 1999 is another name for this regulation.

  The greatest accomplishment of GLBA was to repeal a part of a 1933 statute that had become obsolete. The Glass–Steagall Act was enacted in 1933. The Glass–Steagall Act made it illegal for banks, brokerage firms, and insurance agencies to conduct business together. A bank does not issue insurance or shares at the same time.

  In addition to the above, the GLBA mandates that financial firms report how they store and secure their customers' personal details. Safeguard Rules were adopted by the GLBA and must be implemented. These safeguard principles are spelled out in the legislation. The safeguard laws include, among other items, conducting background checks on staff who may have access to customer details and requiring prospective employees to sign a secrecy pledge.
  - Require secure passwords that are updated regularly.
  - Limit access to private information to those who have a "Need to Know" basis.
  - Require computer screens to lock after a certain period of inactivity.
  - Implement computer and data protection compliance policies.
  - Provide original and occasional compliance instruction to staff, and advise them of the policies on a daily basis.
  - Create compliance protocols for remote workers.
  - Establish procedures for enforcing security breaches and disciplinary action.

- Take precautions to protect data in transit and at rest. Check who has access to this information as well.
- Safely dispose of information.
- Concerning Homeland Security Act (2002)

George W. Bush signed the Homeland Security Act into law in 2002. The Federal Information Security Management Act (FISMA) was included in this act. Following numerous militant threats in the United States, the Homeland Security Act was enacted. The World Trade Center bombing was one of these extremist attacks, as was the sending of anthrax spores to several news agencies and government leaders.

The DHS was created by the Homeland Security Act (DHS). Aside from that, the act served other functions, such as establishing FISMA cyber security regulations. The National Institute of Standards and Technology (NIST) was founded as part of FISMA. The NIST was tasked with creating cyber security standards, protocols, and practices.

The NIST lays out nine measures for achieving FISMA compliance:

- Sort the data you want to keep safe into categories.
- Choose the bare minimum of baseline controls.
- Use a risk management procedure to fine-tune measures.
- In the device protection strategy, document the controls.
- Use effective information management to implement security measures.
- After the security protocols have been implemented, evaluate their efficacy.
- Assess the mission or business case's vulnerability at the department level.
- Give the information system permission to process the data.
- Constantly track the security measures.

3.3   CORPORATE CYBER SECURITY PREPAREDNESS

The Internet and information technology are important tools for small enterprises to expand into new markets and for corporations to increase productivity and production. Businesses, on the other hand, need a cyber defense policy to defend themselves, their clients, and their data from ever-increasing cyber security attacks.

1. Educate staff about the importance of defense.

Establish common security procedures and standards for employees, such as safe passwords and Internet use guidelines that lay out the ramifications of violating the company's cyber security policies. Create policies for managing and securing customer records and other sensitive information.

2. Defend data, devices, and networks from cyber attacks.

Maintain a clean machine: the latest security tools, Web browser, and operating system are the greatest defenses against viruses, ransomware, and other online assaults. Set your antivirus program to run a scan after each download. Other critical security updates should be applied as soon as they are available.

3. Protect your Internet access with a firewall.

   A firewall is a collection of applications that work together to protect data on a private network from unwanted access. Make sure your operating system's firewall is switched on, or use the Internet to find and install free firewall software. If the employees work from home, ensure that their computers are protected by a firewall.

4. Make a strategy for dealing with handheld devices.

   Mobile devices, especially those that hold sensitive data or have access to the business network, may cause significant security and management issues. Require users to password-protect their smartphones, encrypt their data, and install security software to prevent hackers from obtaining information when on public networks. Set up procedures for identifying equipment that has gone missing or has been stolen.

5. Make copies of critical company records and documents as backups.

   On a regular basis, backup the data on both devices. Critical data include word-processing records, computer spreadsheets, directories, accounting files, human resources files, and accounts receivable/payable files. Backup data automatically or at least on a regular basis, as needed, and store copies elsewhere or in the cloud.

6. Build user profiles for each employee and restrict physical access to the machines.

   Unauthorized people should not view or use company machines. Laptops are especially vulnerable to theft or loss, so keep them locked up while not in use. Make sure that each employee has his or her own user account and that good passwords are used. Only trusted IT workers and key personnel should be granted administrative rights.

7. Keep the Wi-Fi networks secure.

   Make sure your workplace's Wi-Fi network is safe, secure, and concealed. To mask your Wi-Fi network, configure your wireless access point or router to not broadcast the network name, also known as the Service Set Identifier (SSID). The router's access is password protected.

8. Use best practices when it comes to credit cards.

   Working with banks or processors to ensure the use of the most trusted and verified software and antifraud programs. Additional protection obligations can be imposed by arrangements with the bank or processor. Separate payment processes from other, less reliable programs, and don't process payments and browse the Internet on the same machine.

9. Restrict employee access to data and records, as well as device installation authority.

   Allowing a single employee access to all computer systems is not a good idea. Employees should only have access to the computer structures that they use for their work, and they should not be allowed to update applications without authorization.

10. Authentication and passwords.

    Employees are required to use special passwords and update them every three months. Consider using multifactor authentication, which needs more detail than just a password to obtain access. Check with your sensitive data vendors, particularly financial institutions, to check if multifactor authentication is available for your account.

3.4 FUTURE OF CYBER LAWS

Political leaders around the world have been struggling to keep up with the fast-paced existence of technical advances since the advent of the Internet in the 1990s. Self-driving vehicles were a far-fetched dream only 20 years ago, and electronic security regulation was centered on the 1990s' main technology focus: computers and the Internet. When we enter a new decade, many people are wondering what shape electronic security legislation will take in the future (Figure 3.2).

Many legislations and legislations relating to cyber security are currently being enacted on a local and national level, raising concerns about future developments. Emerging Technology Laws are becoming more prevalent. The concept of a wired universe was restricted to telephones in the 1950s. Then came the 1990s, when the Internet demonstrated that humanity's technical capabilities far outstripped our wildest expectations. We could never have thought in the 1990s that our culture would shift so quickly online that technology would make our whole lives easier, from operating from home to getting meals delivered to purchasing self-driving vehicles and making artificial intelligence (AI) imitate humans.

Our cyber security regulations can adapt to keep up with the exponential development of emerging modes of technology. Although Congress can be reluctant to enact legislation on each emerging type of technology, passing blanket laws that protect consumers' privacy regardless of technical medium helps to make these changes go more smoothly. While still in their infancy, virtual reality and augmented reality are relatively recent types of technology that pose many concerns with regard to user privacy. Digital reality, for example, is making inroads into the healthcare

FIGURE 3.2 Future of cyber laws.

sector. Would HIPAA and other healthcare-related legislation need to be amended to ensure that patient privacy is covered on these new platforms? What form would such amendments take? Other technologies, such as Alexa and Google Home (automation services), have raised privacy issues, as the microphone settings enable these systems to gather an unprecedented amount of data. Legislation could be needed to limit the volume of data gathered, as well as how the data are processed and exchanged.

3.5   IDENTIFICATION OF THEFT

Identity theft is the stealing of another person's identity or financial information in order to commit fraud using that person's identity, such as making false transactions or purchases. Identity theft may take various forms, but the most frequent outcomes are damage to a victim's credit, money, and reputation. When someone impersonates you and exploits your personal identifying information to perpetrate fraud or gain other financial advantages, this is known as identity theft. Personal identifying information include your complete name, home address, email address, Internet username and passwords, Social Security Number (SSN), driver's license number, passport number, or bank number. Once thieves get access to this information, they will use it to conduct identity theft or sell it on the dark Web. If an identity hacker overhears you reading out your credit card number on the radio, buys your information on the dark Web after it has been exposed in a data breach, or steals it in some other manner, they will have access to your personal information. The next stage in the identity theft process is to use what you've learned.

The following are a few examples of what identity criminals might do with your private details:

- Use your identification details to open new credit cards or lines of credit.
- Make illegal payments using the existing credit and debit cards.
- Demand a refund by filing a tax report with your SSN.
- Seek medical treatment with your health benefits.
- Use your name and financial standing to pass a background check for a job or to rent an apartment.

3.5.1   Data Breaches

A data breach occurs when information is compromised or removed from a device without the owner's consent or permission. A data breach may happen to a small business or a big corporation. Credit card numbers, consumer records, trade secrets, and national security information are examples of private, proprietary, or classified information that may be stolen. The consequences of a data breach may include harm to the target company's image as a result of a suspected "betrayal of confidence." If associated documents are part of the information stolen, victims and their clients could face financial losses. According to the number of data leak events reported between January 2005 and April 2015, the most leaked record category was personally identifiable information, followed by financial data.

Your classified details could be traded on the dark Web after a data breach. A data breach will put the sensitive information of millions of individuals at risk.

The Equifax data hack, for example, compromised the sensitive details of up to 147 million users. Each country or territory has its own set of data breach laws. In certain nations, organizations are also not required to alert authorities in the event of a data loss. Organizations in countries such as the United States, Canada, and France are required to warn impacted people of a data violation until no requirements are met.

Enterprises' Best Practices

- Make appropriate patches to applications and networks. To deter attackers from leveraging bugs in unlatched or obsolete applications, IT administrators should ensure that every program on the network is patched and modified.
- Instruct and execute. Inform the workers about the risks, teach them to recognize social engineering techniques, and develop and/or implement rules for dealing with a threat if one arises.
- Put in place security controls. Create a system for identifying and addressing bugs and risks in the network. Perform compliance checks on a regular basis to ensure that all networks connecting to the company's network are accounted for.
- Plan for the worst-case scenario. Create a disaster response plan that works. Minimize uncertainty in the case of a data breach by having contact information, disclosure plans, and actual mitigating measures on hand. Be sure the workers are aware of this strategy so that they can react appropriately if a violation occurs.
- Keep track of your banking receipts if you're an employee. The first indication that your account has been hacked is the appearance of odd charges on your statement that you did not produce.
- Don't believe everything you see on the surface. Social engineering preys on trust. Be careful and suspicious.
- Be careful what you share on social media. Allowing oneself to get carried away is not a good idea. Keep your profile as secretive as possible if at all feasible.
- Check whether all of your gadgets are secure. These gadgets include laptops, portable computers, and wearables. Ascertain that they are protected by the most recent security software.
- Maintain the safety of your accounts. Use a different email address and password for each user. Using a password manager will make the procedure easier.
- Emails from unknown senders should never be opened. Uninstall emails that seem to be fake without reading them whenever possible. Make sure you know who the sender is and what the email is about before you open it.

3.5.2 Purchase Personal Information from Criminals

Internet criminals have progressed from infecting computers with software to stealing data to forming cybercrime rings that trade different quantities of data for a set fee.

The data may provide terabytes of personal knowledge about Internet users, which the customer would then use if they see fit. These new allegations are very concerning for the protection and security of all Internet users, and analysts have concluded that a coordinated anticybercrime campaign would be required to combat the threat posed by these online criminals.

The information was made public when Finjan, a security company, announced it they had dealt with thousands of these Web providers. The experts at the security conference have reported that Internet piracy is on the rise, and they believe the police can do more to combat it before it becomes uncontrollable. Many of the websites have patient histories, company orders, corporate addresses, and even pension information. The platforms have the potential to influence not only ordinary Internet consumers but also enterprises. "All of this was discovered on a single hacker's server, and we suspect the data was gathered to be traded online." It's even advertised on some forums: "We sell this kind of info, and here's our price list."

Credit card information is becoming more affordable and can be obtained for a few dollars; however, the most lucrative money comes from large company log sheets, which can be purchased for up to $300. UK card losses from telecommunications, Internet, or mail order fraud totaled $290.5 million in 2007, according to recently published banking industry estimates.

The issue is figuring out how to put an end to it. Although Internet users may use spyware blockers and other resources to try to remain secure online, it appears that cyber criminals are still coming up with new ways to obtain information and then sell it. To minimize the risk of your computer being probed for details, make sure your computer is secure and that you have adequate Internet software enabled. Always aim to stick to safer Internet websites and be cautious about what you click on.

The average US Visa and MasterCard card costs about $4 in these markets right now, which is around the same as it was a year earlier. However, Discover and American Express cards distributed in the United States have seen their rates drop by 25% and 15%, respectively. A stolen American Express card cost $7 last year, and a stolen Discover card cost $8. Both are now available for $6.

3.5.3 PHISHING

Phishing is a technique for obtaining personal information through the use of misleading emails and websites. Here's what you need to know about this time-honored yet highly advanced mode of cybercrime (Figure 3.3).

Phishing is a kind of cyber attack in which a fake email is used as a tactic. The goal is to convince the recipient of the email that the letter is something they want or need—such as a bank request or a message from a colleague—and that they may click on a link or download an attachment.

The style of the communication is what distinguishes phishing: the perpetrators pretend to be a trustworthy human of some kind, typically a genuine or convincing real person or an organization with whom the receiver could conduct business. With sophisticated phishing messages and methods, it's one of the oldest kinds of cyber threats, going back to the 1990s. It's also one of the most popular and destructive.

FIGURE 3.3   Phishing.

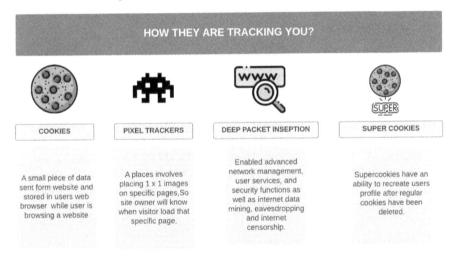

FIGURE 3.4   Spyware (how to track).

3.5.4 SPYWARE

Spyware is a type of malware that collects information about an individual or entity invisibly and then distributes it to third parties. Advertisers or marketing data companies may be among them, which is why spyware is often referred to as "adware." It is installed without the user's knowledge by strategies such as drive-by downloads, Trojans bundled with legal software, or misleading pop-up windows (Figure 3.4).

Spyware collects personal information such as your name, address, surfing habits, desires, hobbies, and downloads from your Internet connection. Other types of spyware can redirect your browser to a different page, make your computer to automatically call or send messages, or serve irritating advertisements even though you are not online. A "keylogger" is a spyware that captures your email, password, or other credentials—an insidious requirement for cybercrime. Unwanted habits and a decrease in system performance are signs of spyware infection. It will use a lot of CPU, disk, and network resources. Application freezing, inability to boot, trouble connecting to the Internet, and server crashes are also typical stability problems. Divergent viewpoints about what constitutes spyware make it a major source of controversy for privacy researchers, who doubt and debate its efficacy. Spyware is almost completely unchecked. Even if the use is legal, these services seldom, if ever, provide a process for the customer to oversee and authorize what information is collected and how it is transmitted. Add to it the fact that spyware uncontrollably consumes computing resources such as bandwidth, processing power, and memory. It's understandable that technology researchers want to avoid and protect against spyware.

3.5.5 IDENTIFY THEFT-MONITORING SERVICES

Swindlers use an ever-evolving variety of tactics and technologies to steal your personal information, making identity fraud more advanced. If effective, a fraudster may use your personal details to open bank accounts, file tax returns, or completely assume someone's name.

It's no longer sufficient to simply dismiss a call from an unknown number. Sophisticated con artists use fake emails (phishing), text messages (smishing), and voicemails (vishing) to persuade you to reveal personal information.

The below are the top identity fraud companies:

> ID Watchdog Platinum Relia Shield Elite Life Lock Ultimate Plus with Norton 360 Privacy; Guard Identity Protection Identity IQ Secure Pro ID Shield Individual 3 Bureaus Experian; Identity Works SM Premium Identity Force Individual ID Shield Family 3 Bureaus ID; Watchdog Platinum Relia Shield Elite Life Lock Ultimate Plus with Norton 360 Life Lock; Ultimate Plus with Norton 360 Life Lock Ultimate Plus.

3.6 ANONYMITY ON INTERNET

The most common concept of anonymity is "not having a name." Simply put, someone is anonymous if their identity is unknown. Being anonymous can be seen as a reduction of a person's responsibility for their acts. Anonymity is linked to anonymity and it is often preferable to avoid having a clear contact with a single person, even though it is often necessary by statute to show an identity before and/or after an event. In the real world, we have many means of identification that are universally accepted, such as the SSN, driver's license, visa, and so on. Pseudonymity is a form of anonymity in which a name other than the author's actual name is shown. The true identity under a pseudonym is often kept very private, and sometimes it is publicly recognized. As opposed to full anonymity, a pseudonym has the benefit of allowing you to know that multiple messages are written by the same person. It is also possible to write a letter to a pseudonym and get responses without knowing who the actual person behind it is. Long conversations between two pseudonyms are likely, even though none of them knows the true identity behind the other's pseudonym. For someone who wishes to remain anonymous, one downside is that mixing information from several messages from the same person can make it easier to figure out who the actual person behind the pseudonym is.

Anonymity isn't something that came out of the Internet. Anonymity and pseudonymity have existed since the dawn of time.

For example, William Shakespeare is almost certainly a pen name, and his actual name is unknown and unlikely to be discovered.

Anonymity has been used for a variety of reasons.

A well-known individual might write messages under a pseudonym if they don't want people's preconceptions of the real author to influence how they see latter. Some individuals may still choose to keep some details of them hidden in order for their messages to be evaluated more objectively. In the past, women often used male pseudonyms, and Jews frequently used pseudonyms in communities where their religion was persecuted. When announcing the findings of a research analysis or discussing particular events, anonymity is often used to protect people's privacy.

In certain nations, there are also rules that protect privacy in some situations.

There are rules in many countries that secure the privacy of tip-offs to newspapers. It is vital that people will report bullying to newspapers, even though they are reliant on the company they are attacking and do not want to expose their true identity. Even though anonymity and pseudonymity are not new to the Internet, the ease at which anonymous and pseudonymous messages can be distributed has improved. On the Internet, anonymity is almost never guaranteed; the perpetrator will still be identified, particularly if the same individual uses the same method to achieve anonymity several times. An individual sends an email or writes a Usenet news article under a false name in the most basic case. Most emails and news applications allow people to choose what names they choose and make no attempt to verify that the user is who they say they are. It is also possible to collect responses and hold conversations using an alias by using Web-based mail services such as Hotmail.

In this case, the anonymity of the customer is not very high. The IP address (physical address) of the device in question is normally registered, as is the host name (logical name). Many users use a temporary IP address allocated to them for a single session to connect to the Internet. However, those numbers are often logged using the ISP (Internet Service Provider), and it is possible to determine who used a certain IP number at a certain time if the ISP cooperates with the investigation. There are also other well-known techniques for breaching secrecy, such as using features on a webpage that convey information without the individual viewing the webpage being aware of it. Some ISPs have a policy of still supporting anonymous users in such searches. As a result, they stop having to make difficult calls about whether to help and when not to help with searches. The email header itself provides a trace of a message's path in the case of email. This detail is not usually displayed to receivers, but most mailers have a command called complete headers that displays it.

3.7 DEFAMATION

Where the expressions, whether spoken or written, are false and inflict injury to another individual, the right to freedom of speech is restricted. Defamation occurs when an oral or written assertion of claimed truth is misleading and causes injury to another individual. The hurt is often financial in nature, affecting a person's ability to make a living, work in a trade, or run for office. Slander is an oral defamatory statement, and libel is a literary defamatory statement. Truth is an utter defense against a claim of defamation because defamation is characterized as an untrue assertion of fact. People have the freedom to voice their views, but they must be cautious about their online contact to prevent defamation claims. Organizations must still be vigilant and ready to respond if they suspect a libelous campaign has been launched against them.

3.8 HATE SPEECH

Hate speech is defined as speech or expression that disparages a person or group of individuals based on their (claimed) membership in a social group characterized by traits such as race, ethnicity, gender, sexual identity, religion, age, physical or mental disability, and others.

Hate speech is described as any form of communication, whether in voice, writing, or actions, that targets or uses derogatory or discriminatory language in relation to an individual or a group based on who they are, such as their faith, ethnicity, nationality, color, sex, descent, gender, or other identification factors. This is often embedded in, and induces, intolerance and hate, and can be demeaning and divisive in certain ways. Rather than banning hate speech outright, international law forbids incitement to bigotry, hatred, and abuse (collectively referred to as "incitement"). Incitement is a particularly dangerous means of expression because it seeks to provoke discrimination, aggression, and brutality, which may contribute to or involve extremism or other atrocity crimes. International law does not compel states to ban hate speech that does not rise to the level of incitement. It is important to emphasize that hate speech can be offensive even though it is not banned. Hate speech has an impact on human rights protection, atrocity crime prevention, terrorism and the underlying spread of violent extremism and counterterrorism, gender-based violence prevention and response, enhanced civilian protection, refugee protection, and the fight against all forms of racial discrimination. As a result, combating hate speech necessitates a concerted response that addresses the root causes and generators of hate speech, as well as the consequences for victims and society at large.

Hate speech includes epithets and slurs, remarks that propagate harmful biases, and speech intended to incite hatred or abuse towards a group. Hate speech may also include nonverbal representations and symbols. Various people and groups have classified the Nazi swastika, the Confederate Battle Flag (of the Confederate States of America), and pornography as hate speech. Hate speech opponents argue that it not only causes psychological suffering and bodily injury to its victims by inciting fear, but it also jeopardizes their social equality. They argue that this is particularly true given the social classes that are often the objects of hate speech have a long history of social marginalization and inequality. As a result, hate speech creates a problem for contemporary western democracies that value both freedom of expression and social inclusion. As a result, in such societies, there is a continuing controversy on when and how hate speech can be restricted or censored.

Under the pretext of freedom of expression, liberals have historically allowed hate speech. Those who share that position acknowledge that hate speech tweets exist, but believe that governmental suppression is a remedy that does more harm than the disease of prejudice. They are worried that a censorship principle will lead to the suppression of other unpopular but lawful speech, such as government criticism, which is necessary for the political health of liberal democracy. They believe that the most effective method to fight hate speech is to demonstrate its falsehood in a free forum of ideas.

3.9 WORKPLACE MONITORING

Employers utilize "employee tracking" methods to monitor their workplaces, as well as the locations and activities of their workers. These methods include employee monitoring devices, time clocks, video surveillance, GPS services, and biometric technology. For example, video surveillance will boost the company's security and efficiency. When a robber is caught on video, shrinkage losses are substantially minimized.

Employee recording and reporting systems help with a variety of other important tasks. Internal fraud is being avoided, employee morale is being assessed, organization services are being appropriately used, and evidence for any future prosecution is being gathered.

- Prediction and flagging systems that attempt to forecast employee characteristics or actions, or that are intended to detect or prevent alleged rule-breaking or fraud. They are marketed as effective compliance tools because they help supplement unfair and unequal processes in job reviews and divide workforces into risk groups based on activity trends.
- Employee biometric and health data gathered through wearables, fitness monitoring software, and biometric timekeeping systems as part of employer-provided health insurance, workplace wellbeing, and automated job change-tracking tools. Tracking nonwork-related behaviors and records, such as health information, can violate worker privacy, create opportunities for harassment, and pose concerns about consent and workers' right to opt out of tracking.
- Employees are managed and success is measured electronically using electronic control and time-tracking. Companies may use these techniques to decentralize and save costs by recruiting independent contractors and maintaining leverage of them through remote management tools, much as regular staff. Advanced time-tracking software can create itemized lists of on-the-job operations, which can be used to encourage wage fraud or enable workers to reduce the amount of time employees are paying to work.
- Continuous data collection enables gamification and algorithmic control of job tasks. Technology may execute management functions such as delivering

FIGURE 3.5 Gamification and algorithmic management.

automatic "nudges" to employees or modifying success benchmarks based on real-time progress, while gamification transforms job processes into competitive, game-like dynamics guided by performance indicators. These policies, on the other hand, will build punitive work conditions in which employees are pressured to reach challenging and changing productivity benchmarks (Figure 3.5).

There is enough evidence to support the conclusion that multiple people spend significant amounts of time at work performing nonwork-related activities. According to a new survey, between 60% and 80% of employees' Internet time is spent on nonwork-related activities. According to another source, staff spend about 4–5 hours a week on personal matters on average. According to a new survey conducted by an IT recruiting company, 54% of businesses have banned the use of social networking platforms such as Facebook and Twitter. IT utilization policy, once established and communicated, sets appropriate conduct limits and allows management to take action against violators. Many companies are monitoring employees to ensure that corporate IT use practices are being implemented, due to the risk of reduced productivity and increased legal liabilities. Many businesses in the United States feel it important to log and monitor employee messages and behaviors on the job, such as phone calls, emails, and Web browsing. Employees are being videotaped on the job in some cases. Random substance tests and psychological testing are still used by several businesses. With a few exceptions, these increasingly prevalent (and, in some cases, intrusive) procedures are completely legal. The Fourth Amendment of the United States Constitution safeguards people from arbitrary government searches and is often used to shield government employees' rights. Staff in the public sector should depend on the "reasonable presumption of privacy" principle created by the *Supreme Court in Katz v. United States* in 1967. The Fourth Amendment, on the other hand, cannot be used to restrict how a private employer handles its workers. As a result, public-sector workers enjoy much more privacy rights than private-sector workers. Although private-sector workers may request legal immunity from an overbearing employer under a variety of state laws, the level of protection varies greatly from state to state. Additionally, state privacy laws favor contractors over workers. Employees, for example, could show that they were in a work place where they had a fair sense of privacy in order to punish a company for violating their privacy rights. As a result, courts often rule against workers who say they were watched when using company devices and file privacy lawsuits. A private corporation can defeat a privacy lawsuit by demonstrating that an employee was given clear warning that his or her email, Internet use, and data on company servers were not private and could be tracked.

3.10   STRATEGIES FOR CONSUMER PROFILING

The only way to gather the insights needed to recognize, segment, and characterize your target demographic is through consumer profiling (also known as "customer profiling"). It means being as close to the customer as possible, so you can meet them in the most effective manner possible. However, rigorous analysis isn't the solution. It is used in high-quality perception. Yeah, there is a distinction. Consumer profiling is the process of extracting value from data in order to learn what you can about your potential customers and the market in which they operate. Leading companies are putting insight in the driver's seat and place customers first in everything they do, from marketing strategy to brand positioning. This process begins with an emphasis on your existing clients, and moves on to the desired potential demographic and target market to ensure you're looking for the right people. If you've got that, it's time to figure out who these people are and how to communicate with them effectively, reflecting on the perfect customer insights that can inspire meaningful innovation.

When people register on websites, complete surveys, fill out paperwork, or join tournaments online, companies freely gather personal information from them. Many organizations often collect information on Web users through the use of cookies, which are text files that are copied to the hard drives of users who access a website in order for the website to recognize them on future visits. Companies also use monitoring tools to monitor surfing patterns and deduce personal tastes and desires on their websites. Since marketers can gather information about users without their prior consent, the use of cookies and monitoring devices is contentious. Cookies allow a website to customize the advertisements and promotions it displays to you after they have been stored on your computer. If the advertiser has agreed to sell using frequency, the marketer understands what commercials have been watched more recently and ensures that they aren't seen again. Some cookies will trace the other websites a user has accessed, enabling marketers to make informed assumptions about the types of advertising that will be more appealing to the user. Companies may collect customer information in the following manner.

3.10.1   Aggregating Consumer Data

Marketing companies combine the data they collect about customers to create databases with massive amounts of information. They want to learn everything they can about their customers, like who they are, what they like, how they act, and what drives them to buy. Marketing agencies offer this information to businesses so that they can adapt their goods and services to the needs of particular customers. Advertisers use the information to better target and entice consumers to respond to their tweets. Buyers should be able to shop more effectively to select items that are well-suited to them, in theory. Sellers should be able to help customize their goods and services to suit the needs of their clients and boost revenue.

3.10.2   COLLECTING DATA FROM WEBSITE VISITS

Marketers use cookies to identify and store information about repeat visits to their websites. The aim is to provide each customer with a unique experience. When a user visits a website, the site will ask the user's machine whether it is okay to store a cookie on the hard drive. If the machine approves, it will be given a unique id and a cookie with that number will be placed on its hard drive. Marketers may use cookies to capture click stream data, which is information collected from tracking a user's online behavior. Three forms of data are collected during a Web browsing session. To begin, "GET" data are collected when one browses the Internet. The second step is to collect "POST" results. POST data are put into blank fields on an accompanying webpage when a consumer joins up for a service, such as Travelocity's service that sends an email when airline prices for preferred destinations change. Third, the marketer keeps track of the details that the customer seeks and views when browsing through some associated webpages. As a result, when someone surfs the Internet, a massive amount of data is created for advertisers and sellers. The four options for limiting or even stopping the deposit of cookies on your hard drive are to change your browser settings so that your machine does not accept cookies, manually remove cookies from your hard drive, download and install a cookie-management program, or use anonymous search programs that do not accept cookies.

3.10.3   PERSONALIZATION SOFTWARE

Personalization software is used by Web advertisers to customize the amount, frequency, and combination of their ad placements, as well as to assess how tourists respond to new content, in addition to using cookies to detect customer data. The aim is to convert first-time site users into paying customers and to increase cross-selling opportunities. Personalization software comes in a variety of forms. When a user visits a website, rules-based personalization software uses market rules tied to consumer desires or online history to decide the most suitable page views and product details to present. For example, if you use a website to book plane tickets to a common vacation destination, rules-based software can ensure that you see rental car advertising.

Collaborative filtering allows product choices dependent on what other consumers with common shopping preferences have bought. If you purchased a book by Dean Koontz, for example, a company could suggest Stephen King books to you based on the fact that a large number of other buyers bought books by both writers.

Another kind of personalization program is demographic filtering. It generates product suggestions by integrating click stream data and user-supplied data with demographic data provided with user zip codes to create product recommendations. It has amassed a massive database of tens of millions of users, each with their own unique user ID. Microsoft has now created a database-based technology that allows advertisers to tailor one advertisement to men and another to women. Ad-selection criteria may provide additional details such as age and position.

Contextual commerce, another form of personalization tech, links product deals and other e-commerce services to particular information a customer might see in an online news article. For example, when reading a story about white-water rafting, you may be given a discount on rafting equipment or a coupon for a West Virginia white-water rafting holiday.

Sensitive information, such as names, addresses, and SSNs, can only be collected if individuals supply it. Companies cannot access individual Web surfers who visit their pages without this knowledge. Cookies collect anonymous information about a user's Web surfing as long as the network advertiser does not connect the data to personal information. A website user may use personal information provided by a tourist to discover additional personal information that the visitor does not wish to share. For example, a name and address may be used to find a phone number, which may contribute to the collection of much more sensitive information. All of this knowledge is highly useful to a website user who is attempting to establish a relationship with guests and convert them into consumers. This information may be used to initiate correspondence or sold to other companies for which the operator has marketing agreements.

3.10.4   CONSUMER DATA PRIVACY

Consumer privacy, also known as customer privacy, refers to how confidential personal information collected by users is handled and protected during daily transactions. User data protection has become an increasing concern as the Internet has developed into a commercial medium.

3.10.4.1   General Data Protection Regulation

The General Data Protection Regulation (GDPR) is a data and privacy protection regulation. It extends to all inhabitants of the European Union and the European Economic Area, giving them more data ownership and power, as well as more data-collection rights.

It's worth noting that, at a high level, GDPR refers to companies doing business with EU residents who are based outside of the EU and EEA.

The CCPA is close to GDPR in that it gives users greater power, protection, and confidentiality of their personal information. However, as the name implies, it refers to California residents and anybody who can sell or gather their info.

The GDPR and CCPA, in my mind, are just the tip of the iceberg, and we will eventually see either a nationwide level in data and privacy rights legislation or more states joining in. Prior to the CCPA, Massachusetts had the strictest privacy laws in the country; now, other states are following suit.

According to a new survey, if a website isn't safe, 84% of consumers will abandon an online order. The EU approved the historic GDPR in 2018 to expand data privacy protections, and California declared earlier this year that there will be no delay in implementing the newly passed "CCPA" due to the pandemic.

Consumers will have greater awareness and discretion of what sensitive information is stored, how it is sold, and how its protection is ensured as a result of this legislation. Marketers have taken a number of measures to conform to the new rules, including opt-in checkboxes on online forms that ask for permission before connecting users to a mailing list, making privacy statements and disclosures easily accessible, and having special webpages where users can make requests to view, alter, or uninstall their personal details.

Although both B2B and B2C advertisers have taken special precautions for some time to ensure that prospects have given their permission to be targeted, the new environment has made maintaining sufficient safeguards even more difficult. Many nondigital enterprises have been driven to transition to Internet activities as a result of the pandemic, and many of these businesses lack remote job policies and protocols for handling confidential data outside of the workplace.

The Indian government is expected to pass a Personal Data Protection Bill (DPB) that will regulate the collection, encoding, storage, use, transfer, protection, and dissemination of personal data about Indian citizens. DPB is a significant achievement for global managers, despite its geographical presence. By 2022, India's digital economy is estimated to be worth $1 trillion, attracting a slew of international players who will have to cooperate with the DPB.

Instead of adopting China's isolationist regulatory system, which prohibits foreign players such as Facebook and Google from working within its borders, India has adopted the EU's GDPR, which allows global Internet firms to do business under certain conditions.

3.11 ELECTRONIC DISCOVERY

Electronic discovery (also known as e-discovery) is the electronic aspect of locating, gathering, and producing electronically stored information (ESI) in response to a production order in a lawsuit or prosecution. Emails, notes, presentations, directories, voicemail, audio and video recordings, social media, and webpages are all examples of ESI.

Because of the vast amount of electronic data generated and processed, e-discovery processes and technology are often complex. Electronic records, unlike hardcopy evidence, are more complex and often include metadata such as time and date stamps, author and recipient information, and file assets. To avoid charges of spoliation or tampering with documentation later in the proceedings, it is essential to preserve the original material and metadata for electronically stored records.

Potentially valid records (including all electronic and hard-copy materials) are put in a legal lock after evidence is found by both parties of a case. This means they cannot be changed, removed, erased, or otherwise damaged. Data that may be important are gathered, then compiled, indexed, and stored in a database. At this stage, the data are processed to cull or separate the records and emails that are simply irrelevant. The information is then stored in a safe environment and made available to reviewers who code the records according to their legal significance (Figure 3.6).

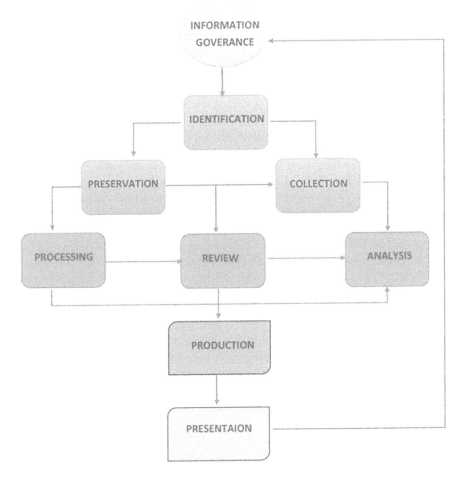

FIGURE 3.6   Electronic discovery reference model (EDRM).

For output, sensitive documents are often translated to a static file like TIFF or PDF, allowing for the redaction of privileged and nonrelevant content. For e-discovery, the use of computer-aided analysis predictive coding and other computational tools decreases the amount of records that need to be reviewed by lawyers and helps the legal staff to select the documents that are reviewed. The number of records is reduced, which saves time and money. The end aim of e-Discovery is to provide a central amount of defensible evidence for litigation.

The EDRM is a visual representation of the different stages of the e-Discovery process. It's broken down into nine stages that explain how e-Discovery works throughout an inquiry. It's important to note that these stages don't always happen in the sequence indicated. The EDRM represents an iterative process that may be completed in any sequence.

Many legal departments use it to direct their e-Discovery processes since it is very well-known and commonly used. This isn't, though, a prescriptive paradigm designed to teach e-Discovery practitioners how they can handle the process. "The EDRM diagram reflects a philosophical view of the e-discovery process, not a literal, linear, or waterfall model," the developers claim. Some but not all of the steps outlined in the diagram may be completed, or the steps may be completed in a different order than shown in figure."

3.12   CHILDREN'S INTERNET PROTECTION ACT

As a condition of receiving such federal grants, such as E-rate funds, the Children's Internet Protection Act (CIPA) mandates K-12 schools and libraries to use Internet filtering and introduce other steps to shield children from unsafe online content. It was signed into law on December 21, 2000, and the US Supreme Court ruled on June 23, 2003, that it was lawful. CIPA is one of the bills introduced by the US Congress to restrict children's Internet exposure to pornography and sexual videos. The Communications Decency Act and the Child Online Protection Act, two previous efforts by Congress to regulate obscene Internet content, were both declared unconstitutional by the US Supreme Court on First Amendment grounds.

On December 15, 2000, the CIPA and the Neighborhood Internet Protection Act were enacted by Congress as part of a new budget package (H.R. 4577). On December 21, 2000, President Clinton signed the bill into law (Public Law 106–554). The Acts impose limitations on the usage of funds provided by the Library Services and Technology Act, Title III of the Elementary and Secondary Education Act, and the E-rate Universal Service Discount Scheme. These limitations come in the form of Internet safety policies and technologies that prevent or filter certain types of content from being viewed using the Internet.

CIPA's requirements

CIPA-affected schools and libraries will not be eligible for E-rate savings until they certify that they have an Internet safety strategy that incorporates security prevention initiatives. Pictures that are

a. obscene;
b. child pornography; or
c. dangerous to minors must be blocked or filtered from being seen on the Internet (for computers that are accessed by minors).

Schools and libraries must have fair warning and schedule at least one public hearing or consultation before implementing this Internet protection program. CIPA-affected schools must meet two additional qualification requirements:

1. Their Internet protection measures must include tracking minors' online activities; and
2. They must provide for teaching minors about proper online activity, such as engaging with others on social networking networks and in chat rooms, as well as cyber bullying tolerance and response, as required by the Protecting Children in the Twenty-First Century Act.

CIPA mandates that schools and libraries develop and execute an Internet safety strategy:

- Minors' access to inappropriate material on the Internet;
- Minors' safety and security by using email, chat rooms, and other forms of direct electronic communications; and
- Unauthorized access, such as "hacking," and other illegal internet actions by children. Schools and libraries must declare their compliance with CIPA before obtaining E-rate funding.
- During the usage of an adult, an authorized person may deactivate the blocking or filtering mechanism to enable access for bona genuine analysis or other legal reasons.
- CIPA does not mandate the monitoring of Internet use by minors or adults.

3.13   INTERNET FILTERING

To comprehend what Internet filtering entails, we must first comprehend what filter entails, and then we can comprehend all meanings. Filter is a program or a piece of pass-through code that examines an input or output request to see whether it meets standards before processing or forwarding it. Filter accepts data as input, analyzes it, and makes a decision before passing it over to another application. In most cases, a filter does not have its own input or output operation. In such data, a filter is used to insert or delete headers or control characters. The filter compares the contents of a webpage to a collection of guidelines given by someone who has already installed internet filter software. The request for that webpage would not be completed if the contents of that webpage are in violation of the law. Furthermore, filtering can be allowed on:

- Webpages
- Keywords and accounts
- Keyword filters
- Directories and extensions
- Block URL filters
- Filters for applications and archives
- Filters for Web objects
- Filters for flash and cookies

Filters come in a variety of shapes and sizes, including:

Hardware: It makes use of a hardware interface to link the machine to the Internet. It's possible that it's a cable modem or a DSL router. It is contingent on the kind of broadband service you buy.

Software: It comes with a device program that is always the most stable and has the highest degree of security.

DNS: A DNS server, such as Open DNS, is used to provide filtering. This style of filter offers a cost-free alternative with many of the benefits of a hardware solution. These benefits would filter all electronics in the household, such as laptops, mobile phones, and televisions.

Filtering by proxy is one of the services provided by ISPs. It will use the Internet to filter your content. Just make certain that the machine is set up to use proxy.

Today, one of the most important security measures is Internet filtering. Content management software or site-filtering software that blocks unauthorized content or materials over the Internet is referred to as Internet filtering. It will block pop-up pages, advertisements, inappropriate content or unwelcome websites, viruses, file transfers, and chat rooms. So, if a webpage has been censored, the Internet filter program cannot complete the blocked webpage search. Internet filters are used to keep track of what people are able to read on the Internet. It will, for example, assist parents in monitoring their children's Internet use. Since children are vulnerable to a variety of threats, such as fake websites and malware files infected with viruses, they should be protected. However, the key reason that parents watch their children when they are online is that children can spend their time playing video games rather than doing schoolwork. Guardians can pick and classify which websites their children can access using Internet filtering software. Companies may also restrict access to pages that they do not want their workers to use at work. Your filtering is determined by the tools you use and the controls you have. You can set up various levels of controls with filtering tools. So that you can limit your children's access to some content, allow your teenagers more access, and give guardians free Internet access.

Internet filtering secondary benefits

Internet filtering can improve office efficiency by restricting links to nonwork-related websites, in addition to reducing the possibility of malware and ransomware. Millions of webpages are divided into more than 50 groups (abortion, adult content, alternative views, alcohol, etc.) by Internet filtering appliances, which can be blocked with the click of a button (Figure 3.7).

Users may also be prevented from viewing objectionable content by blocking those types of websites. This will protect consumers, teachers, and diners—as well as staff—from being subjected to derogatory Web content if a company has a freely available Internet service. It would remove several HR problems associated with harassment and discrimination in the workplace.

3.14   ADVANCE SURVEILLANCE TECHNOLOGY

Surveillance cameras, facial recognition technologies, and satellite-based devices that can detect a person's physical location, among other advancements in information technology, offer groundbreaking new data-gathering capabilities. However, these advancements can jeopardize individual privacy and confuse the question of how much details about people's private lives can be collected. Surveillance technology, which includes a broad range of tools used to monitor people's behaviors and interactions, is rapidly evolving and becoming more readily accessible in the general market. This pattern may be attributed to the terrorist concerns, as well as surveillance equipment manufacturers' willingness to develop the technology at lower prices. Although the military and government intelligence services have long used

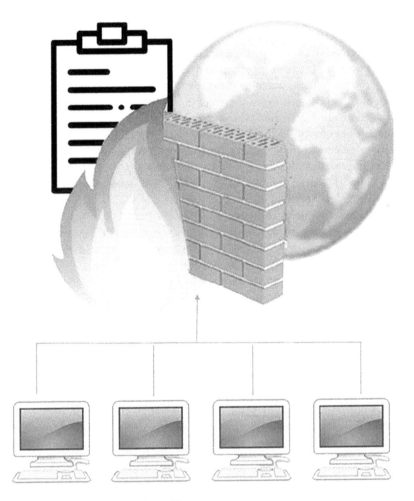

FIGURE 3.7   Internet filtering (firewall).

surveillance technology, its use by law enforcement and even private citizens in fields that are not of national security interest is fresh, and it poses privacy and evidentiary questions that are yet to be resolved by US courts. The following sections explain some of the innovations.

3.14.1   CAMERA SURVEILLANCE

Surveillance cameras are cameras that are used to keep an eye on a particular area. They are often connected to a recording system or an IP network, and they may be monitored by a security guard or law enforcement official. To stop corruption and terrorism attacks, surveillance cameras are used in big cities around the world. Critics argue that such scrutiny violates civil rights, and they are worried about the cost of the facilities and personnel needed to track the video feeds. Supporters of surveillance cameras point to observational evidence that the cameras are useful in stopping violence and extremism. They will provide examples of how cameras aided in the investigation of crimes by correlating witness statements and assisting in the tracking down of criminals. In the United Kingdom, there are 4.2 million closed circuit television (CCTV) cameras in use, which equates to one CCTV camera for every 14 people. For example, China has 2.75 million cameras or one camera for every 472,000 people. During the 2012 Olympics, the number of cameras in London was significantly increased, and a device called DYVINE allows all London CCTV cameras to be tracked and operated from New Scotland Yard.

Applications of security cameras

Safety cameras are used in almost every industry around the world due to their wide variety of designs and functionality. Theft and vandalism prevention, traffic and weather control, and other uses are possible.

An IP camera can be mounted in almost any location by integrating advanced camera functionality with rugged outdoor housings or discreet camera architecture. This makes them suitable for surveillance inside and outside of businesses and workplaces, filming in remote areas, and recording high-resolution footage to detect suspected individuals.

- Parking Lot Surveillance's Advantages

    Prevent Vehicle Theft—Many criminals see parking lots as gold fields, with miles and miles of vehicles and personal items to plunder. Seen surveillance cameras will prevent offenders from breaking into cars on your lot and assist law enforcement in detecting suspects who do break into cars on your lot. Shopping carts, outdoor exhibits, tables and benches, landscaping, and other store property can all be protected with cameras.

    Increase Public Safety—For many companies, customers are the most important factor. Keeping your consumers happy and providing them with a sense of security and peace of mind as they shop at your store builds confidence and loyalty.

    Lower your liability—In your parking lot, an accident is almost certain to occur. If it's a traffic accident, a slip, or an errant shopping cart denting cars, video recording is a cost-effective way to prevent expensive insurance charges and even lawsuits.

    Identify vehicles that aren't charged or aren't allowed—Drivers can choose to exit the parking lot without paying, similar to how they can "dine-and-dash" in restaurants. Although most parking lots have attendants, it's almost impossible to guarantee that all areas of the lot are guarded by workers. Surveillance cameras will record license plate numbers, helping you to determine whether or not to press charges or bar your customer from using your parking lot again.

3.14.2 GPS Chips

The US Air Force created the Global Positioning System (GPS) to aid them in combat operations. GPS has had a huge effect on both positioning, navigation, and tracking applications around the globe, and it has grown to the point that it is used in virtually every part of our lives over the last few decades. Users can get instantaneous location and time data from GPS chips and modules anywhere on the planet. GPS receivers obtain the data and analyze it to determine the receiver's time and location. GPS works under the following principle: signals from at least three satellites are used to measure the location of the receiver on Earth (latitude and longitude), and a fourth satellite is used to calculate the altitude in relation to the Earth's surface. The details on where the satellite is in the atmosphere (orbital data) and the time (timestamp) when the signal is received are used in satellite signals. This information causes the receiver's computer software to measure the distance to the satellite.

GPS chips are being installed in a wide range of products, from cars to cell-phones, to precisely identify consumers. The Federal Communications Commission has requested mobile phone providers to adopt mechanisms for locating customers so that police, ambulance, and emergency services can be deployed accurately to help 911 callers. Personal digital assistants, desktop computers, cars, and boats may also use similar location-tracking technologies. Parents will install one of these chips in their teen's car and watch the vehicle's location using apps. Banks, supermarkets, and airlines are all keen to get real-time access to consumer location data, and they've already come up with a list of new services they want to provide, such as delivering digital coupons to customers who are near a shop, offering the location of the closest ATM, and keeping travelers up to date on flight and hotel information. Airlines are considering using wireless technology to allow passengers to check in for flights when near the gate, as well as to monitor when and person passes through the gate. Businesses say that they can protect wireless users' privacy by allowing them to opt in or out of marketing services depending on their location data. Wireless spamming is a distinct possibility—while walking down the street, a person might receive wireless advertisements, announcements for nearby restaurants, and shopping advice. Another problem is that the information could be used to track down people at any time or determine where they were at any given moment. When using a mobile phone, the possibility of revealing one's location.

3.15 LEGAL DEVELOPMENTS IN CYBER LAWS

In a nutshell, cybercrime is any illegal activity in which a device is used as a weapon, a target, or both. Classic criminal acts such as stealing, bribery, forgery, slander, and mischief, which are covered under the Indian Penal Code, can be used in cyber-crimes. The Information Technology Act of 2000 addresses a variety of new age violations that have arisen as a result of electronic misuse.

Cybercrime can be classified in the following two ways.

- Use a device to strike other computers (the computer as a target). For example, hacking, virus/worm attacks, DOS attacks, and so on.

- Using a computer as a weapon entails using a computer to conduct real-world crimes. Only a few instances include cyber terrorism, IPR breaches, credit card frauds, EFT scams, pornography, and so on.

The phrase "cyber law" refers to the legal issues surrounding the use of communications technology, particularly "cyberspace" or the Internet. It is more of a nexus of many legal domains, such as intellectual property, anonymity, freedom of expression, and power, than a distinct field of law such as property or contract. Cyber law, on the other hand, aims at reconciling the issues raised by human interaction on the Internet with the existing legal system that regulates the physical world.

The creators of the Internet had no clue when it was initially established that it would evolve into an all-encompassing revolution that might be exploited for criminal purposes and would need monitoring. These days, a lot of disturbing things have been happening online. Because of the anonymity of the Internet, it is possible to engage in a broad variety of unlawful actions while remaining anonymous, and individuals in positions of intelligence have been severely abusing this aspect of the Internet to conduct criminal operations in cyberspace. The IT Act of 2000 was enacted to update current laws and provide remedies to cybercrime. That law is required so that consumers may use credit cards to make online transactions without fear of fraud. The Act establishes a much-needed legislative framework to guarantee that electronic evidence retains its legal impact, legitimacy, and enforceability.

In view of the increasing number of transactions and communication performed through electronic records, the Act seeks to enable government agencies to recognize the filing, development, and storage of official documents in digital format. The Act also provides a legal framework for the verification and origination of electronic documents and communications using digital signatures.

- In terms of e-commerce in India, the IT Act of 2000 and its regulations offer a lot of potential. To begin, these regulations will have implications for e-businesses in that email will now be recognized as a genuine and legal method of communication in our nation, capable of being produced, and accepted in a court of law.
- Businesses will now be able to conduct electronic commerce using the Act's regulatory infrastructure.
- The Act establishes the formal status and sanction of digital signatures.
- The Act makes it possible for corporations to become Certifying Authorities for the purposes of issuing Digital Signature Certificates.
- The Act also mandates that the government send out alerts over the Internet, ushering in a new era of e-governance.
- The Act enables companies to submit any form, proposal, or other document to any institution, jurisdiction, organization, or agency owned or managed by the relevant government in electronic form, using any electronic form specified by the applicable government.

The IT Act also addresses the critical security issues that are critical to the success of electronic transactions. Protected digital signatures, which would have to go through a set of verification processes that the government would define at a later point, now have a legal meaning according to the Act.

- Under the Information Technology Act of 2000, businesses will now have a contractual solution if anyone hacks into their operating networks or network and causes disruption or copies records. The Act provides for punitive damages of not more than Rs. 1 crore as a relief.

3.16   FEDERAL TRADE COMMISSION ACT SECTION 5

The United States does not actually have a robust data retention regulation comparable to the European Union's Data Protection Directive. Instead, the FTC uses its limited jurisdiction under Section 5 of the Federal Trade Commission Act (15 U.S. Code 45) to complement industry-specific regulations.

The FTC has the power to prosecute "unfair or misleading actions or activities in or influencing commerce," according to Section 5. The FTC has used Section 5 to investigate "unfair" or "deceptive" data protection and privacy policies, focusing on the "deception" component rather than allegations of "unfairness."

To prevent a Section 5 breach, the FTC recommends that enterprises implement a "privacy by design" policy, simplify data options for businesses and customers, and increase openness of activities. The privacy by design philosophy allows businesses to think about future privacy and data protection concerns early in the production of their organization, product, or service. Furthermore, the baseline privacy by design guideline is supported by two applied principles: Data confidentiality, fair collection limits, good preservation and recycling procedures, and data integrity can all be part of a company's privacy policies. Throughout the life cycle of their goods and services, businesses should maintain robust data-processing procedures.

General Counsel must consider privacy by design in order to educate businesses about how to ensure that their procedures and goods or services comply with regulatory standards, posted privacy policy, product specifications, and consumer preferences. The principle is constructive rather than reactive in preserving private information, with the assumption that it is more difficult to "fix architectural shortcomings after rollout." The FTC also advises companies and policymakers to implement "simplified choice processes" that allow users "more direct leverage" over their privacy and make data policies more transparent.

Many computer security and privacy proceedings, according to Jennifer Woods of Clark Hill PLC's intellectual property division, are not exclusively dependent on a Section 5 breach, but could also be linked to other FTC-mandated laws such as the Fair Credit Reporting Act and the Children's Online Privacy Protection Act.

The FTC is empowered to prosecute privacy and data protection breaches fewer than 60 separate sets of regulations, codes, and guidelines. The United States Code contains laws, whereas Title 16 of the Code of Federal Regulations contains most codes.

Cyber threats and how lawyers can help in cyber security, role of lawyers, and why do lawyers participate

As shown by the plethora of cyber security rules, legislation, and policies, the topic has evolved into a legal as well as a technological issue. Most businesses used to assume that data defense should be handled only by IT and risk management. Although some people still believe this, the debate has changed from whether or not lawyers should be interested in a company's cyber defense activities to whether lawyers should be involved. Lawyers are ideally qualified to apply applicable legislation to a company's facts and conditions, determine enforcement, and advise decision-making for firms' data protection efforts in relation to the law. Indeed, the majority of corporate lawyers surveyed for this analysis are interested in their company's cyber defense activities to some degree, with the majority stating that they are "moderately" involved.

3.17   PUBLIC CONCERNS ABOUT SECURITY

For Americans, privacy conjures up a slew of ideas, some of which are rooted in common conceptions of civil rights, while others are motivated by fears over Internet monitoring and the coming age of "big data." Although Americans' feelings about privacy are varied, according to a recent Pew Research Center poll, the majority of adults believe their privacy is being threatened on key dimensions such as the protection of their personal details and their right to maintain confidentiality.

If people in the United States are asked what comes to mind when they hear the word "privacy," they react in predictable ways. They offer a lot of weight to the notion that privacy extends to personal material—their space, their "things," their solitude, and, most notably, their "freedom," as seen in the word cloud above. When responses are clustered into themes, the largest block of responses links to concepts of security, safety, and health, regardless of the frequency of individual terms. When many people think about anonymity, confidentiality and having information "secret" come to mind first.

The majority of people are mindful of the government's attempts to regulate connectivity. More than a year after contractor Edward Snowden leaked documents about systematic government spying by the National Security Agency, the public continues to be shocked by the findings. About 43% of adults have heard "a lot" and 44% have heard "a bit" about "the government gathering information about mobile calls, emails, and other Internet messages as part of attempts to detect criminal violence." Only 5% of adults in our survey said they had known "almost nothing" about these services.

3.18   PREVENTIVE MEASURES TO AVOID
SPEAR PHISHING ATTACKS

Spear phishing is the practice of impersonating a trustworthy sender and sending emails to specific and well-researched targets. The goal is to infect computers with malware or convince victims to provide personal data or valuables. "Spear phishing is a campaign that was intentionally developed by a threat agent with the intention of penetrating one organization, where they can really study names and positions within a company," Higbee continues.

Spear phishing attacks are more complex than mass phishing attacks, which often involve using automated off-the-shelf kits to collect passwords en masse using fake log-in pages for popular banking or email services, or spreading ransomware or crypto mining malware. To steal personal information or proprietary intellectual property, or merely hack payment processes, some targeted campaigns use documents containing ransomware or ties to password theft pages. Others ignore malicious payloads altogether, instead relying on social engineering to take control of systems that result in a limited number of massive payouts made by a single or sequence of bank transactions.

Preventive Actions

1. Use AI to your benefit.

   Find a solution that can detect and block spear phishing threats, such as BEC and brand impersonation, even though they don't have malicious links or attachments. Machine learning software can examine an organization's communication patterns and detect any deviations that could indicate an attack.

2. Don't depend on conventional security measures alone.

   Traditional email protection, which relies on blacklists to detect spear phishing and brand impersonation, may not be able to detect zero-day connections, which are common in many attacks.

3. Set up anti-account-takeover safeguards.

   To prevent further spear phishing attacks from coming from such accounts, look for software that use AI to detect when accounts have been hacked.

4. Make DMARC verification and reporting a reality.

   Domain spoofing and brand hijacking are popular tactics used in impersonation attacks, and DMARC authentication can help deter them.

5. Make multifactor authentication a priority.

   Multifactor authentication is an effective security mechanism that provides another layer of security to a standard username and password.

6. Teach employees how to spot and report threats.

   Any cyber awareness preparation should include identifying and monitoring spear phishing attacks. Users may be trained to recognize phishing texts, voicemails, and text messages using phishing simulations. Businesses can also have processes in place to validate all financial inquiries received by email.

7. Conduct inquiries ahead of time.

   Employees cannot often notice or track spear phishing attacks because they are too customizable. Companies can run routine scans for emails containing material that is considered to be common with hackers, such as subject lines relating to password changes.

8. Increase the likelihood of data loss avoidance.

   Combine security technology with business policies to guarantee that emails containing sensitive or private information are stopped and do not escape the company.

9. File sandboxing inbound.

   Install a solution that verifies the security of an emailed connection when a user clicks on it. This guards against a modern phishing technique used by cybercriminals, as I've seen. To get through the organization's email encryption, bad guys give their targets a brand new URL in an email. The other technique is to insert malicious code into the website immediately after the email URL is sent. Any ordinary spam filter will miss this URL.

10. Analyze and inspect the Web traffic in real time.

First, at your portal, block malicious URLs from reaching your users' corporate inboxes. Even if your corporate email has inbound email sandboxing, some users can click on a malicious connection from a personal email address, such as Gmail. Your company's email spear phishing security won't be able to see the traffic in that situation. Bottom line: the Web protection gateway must be insightful, monitor content in real time, and block malware with a 98% success rate.

11. Employee behavior: the importance of the human aspect cannot be overstated. Adopt an executive testing program and continue to educate employees. Employee education and security training aren't the end results; behavior change is.

REVIEW QUESTIONS

1. Why are companies increasingly utilizing workplace surveillance, and how are they doing it?
2. Write short notes on Defamation and Hate speech.
3. What advantages does consumer profiling provide you as a customer? Do these advantages outweigh the loss of privacy?
4. Write short notes on (a) Children Internet Protection Act and (CIPA) (b) Internet Censorship
5. Discuss key privacy and anonymity issues.
6. "The right of privacy is crucially important for every individual." Comment on the statement.
7. Write short note on identify theft and explain the term phishing.
8. What do you mean by advance surveillance technology? Explain the two surveillance technology in brief.
9. What is identity theft and how do identity thieves operate?

MULTIPLE CHOICE QUESTIONS (MCQs)

1. Which of the following is not a proper aspect of user integration?
   a. employee's authentication
   b. physical authorization
   c. access control
   d. representing users in the database

2. It is very important to block unknown, strange and _____ within the corporate network.
   a. infected sites
   b. programs
   c. unwanted files
   d. important folders

3. A _____ takes over your system's browser settings and the attack will redirect the websites you visit some other websites of its preference.
   a. MiTM
   b. browser hacking
   c. browser hijacker
   d. cookie stealing

4. _____ has become a popular attack since last few years, and the attacker targets board members, high-ranked officials, and managing committee members of an organization.
   a. spyware
   b. ransomware
   c. adware
   d. shareware

5. _____ is the technique to obtain permission from a company for using manufacturing and selling one or more products within a definite market area.
   a. algorithm licensing
   b. code licensing
   c. item licensing
   d. product licensing

6. Which one of the following does not come under security measures for cloud in firms?
   a. firewall
   b. antivirus
   c. load balancer
   d. encryption

7. It is important to limit _____ to all data and information as well as limit the authority for installing software.
   a. workload
   b. employee access
   c. admin permission
   d. installing unwanted apps

8. One must isolate payment systems and payment processes from those computers that you think are used by _____ or may contain _____
   a. strangers, keyloggers
   b. strangers, antivirus
   c. unknown, firewalls
   d. unknown, antivirus

9. _____ is a technique for returning encrypted material, often known as cypher text, to its original form, plain text.
   a. decryption
   b. cryptanalysis
   c. encryption
   d. reverse engineering

10. **Which of the following is not cryptography's main goal?**
    a. authentication
    b. nonrepudiation
    c. data redundancy
    d. confidentiality

ANSWERS TO MCQs

Q1: b, Q2: a, Q3: c, Q4: b, Q5: d, Q6: b, Q7: b, Q8: a, Q9: a, Q10: a

4 Intellectual Property

LEARNING OBJECTIVES

- To get a basic understanding of the laws governing confidential information, copyright, patents, designs, trademarks, and unfair competition.
- Should be able to identify, apply, and assess principles of law relating to each of these areas of intellectual property (IP).
- To understand the legal and practical steps required to ensure that IP rights (IPR) remain valid and enforceable.
- To understand current and emerging issues relating to the IP protection.

4.1 INTRODUCTION TO INTELLECTUAL PROPERTY

India is one of United Kingdom's most significant international markets. If you want to do business in India, or if you currently do business there, you must understand how to use, protect, and uphold the rights to IP that you or your company owns. Here we discuss IP in general and provide advice on how to apply these concepts in the Indian market. It explains how to deal with IP infringement in India, gives guidance on how to deal with it efficiently, and provides links to additional resources.

IPR arising from intellectual activity in the areas of business, research, literature, and the arts are known as IPR. IPR, like any other property, may be exchanged, meaning they can be bought, leased, or acquired. Nonexhaustible and intangible commodities exist. Businesses are diligent in finding and protecting IP since it is so important in today's knowledge-based economy. It is essential for a company's duty to extract income from IP and to prevent others from doing so. IP may take many different forms and sizes. Although IP is an intangible asset, it may be considerably more costly than a company's physical assets. Because IP may offer a strategic advantage, it is carefully guarded and protected by the companies that possess it.

4.1.1 BLOCK DIAGRAM OF IPR

IP is a creation of human intelligence, and the privileges given to it enable its holders to benefit from the fruits of the intellectual endeavor by granting them a monopoly. Such a privilege is not always a common right, and it must be recognized by law.

The following statutes recognize IP rights in India:

- The Patents Act, 1970;
- The Trade Marks Act, 1999;
- The Copyright Act, 1957;
- The Designs Act, 2000;

- The Geographical Indications of Goods (Registration & Protection) Act, 1999;
- The Semiconductor Integrated Circuits Layout Design Act, 2000;
- The Biological Diversity Act, 2002.

IPRs are important in every industry and have formed the basis for key investment choices. Since IPRs are proprietary rights, striking a healthy mix of the needs of innovators and the interests of society as a whole is often a challenge. Another critical consideration is providing an appropriate regulatory system to safeguard innovators' rights and encourage trust for the sake of their own safety IP, which would lead to further creativity.

Due to the vast number of judges, differing degrees of judicial officials' familiarity with IPR issues, and different methods of procedure, IPR litigation in India is very complex. As a result, some courts have surpassed others as favored venues. The majority of infringement suits have been filed in Delhi and Mumbai, especially in their High Courts. Although these jurisdictions have been chosen because of their courts' vast expertise and knowledge of IPR issues, their high number of cases is also attributed to the fact that India's capital is Delhi, and the country's commercial capital is Mumbai. As a consequence, massive trading conflicts are inevitable.

With the exponential growth of cases involving IPR laws—such as trademarks, trade secrets, copyrights, and design laws—the Delhi High Court and the Bombay High Court have received a total of 727 cases in the last year. Trademarks were used in 531 of the cases, followed by copyrights (157), compositions (14), and patents (14). Ex parte injunctions were issued in approximately 532 cases, and permanent injunctions were granted in approximately 661 cases. In the previous year, only 92 lawsuits were resolved. With the goal of simplifying and expediting commercial litigation, including IP issues, the Indian Legislature created commercial courts in India by passing a new Act in 2015, which entered into effect in 2016. Commercial courts and commercial divisions within High Courts of Ordinary Civil Authority have been established at the district level as a result of this legislation. Commercial lawsuits, which involve cases involving IP, have been given a special treatment under this new law. Given the high number and frequently slow progress of litigation, the new Act establishes strict deadlines for all phases of a lawsuit, leaving no space for delays. If a party believes that its adversary has no realistic possibility of winning in or defending an argument and that recording testimony will be unnecessary, it may file a motion for summary judgment under the said Act. The aim is to obtain oral arguments heard and concluded within six months of the parties finishing their paper reviews. This technique is designed to ensure that suits are disposed of quickly.

Apart from reforms in the legislation to increase the pace of courts, the government has taken steps to train judicial officers so that IPR cases can be resolved quickly. The government's Cell for IPR Promotion and Management (CIPAM) has been organizing training and sensitization program on IPRs for District Court and High court judges in collaboration with WIPO and the National Judicial Academy (NJA) of India. The NJA of Bhopal hosted two colloquiums on commercial law for High Court judges. In January 2017, CIPAM, in collaboration with NJA, arranged a Colloquium on Commercial Laws for High Court Justices, which included a forum on IP rights.

With courts that have a better grasp of IPR rules, India is starting to be recognized as a pro-innovator jurisdiction. Furthermore, with a proper legal system in place and no prejudice based on the ethnicity of the individuals claiming their IP rights, India has seen a massive rise in innovators asserting their IP rights (Figure 4.1).

This is one facet of India's litigation patterns, especially in IPR cases. On the other hand, after an ex parte warrant is issued, it takes a long time to lift the injunction and for the trial to be completed. Furthermore, after an injunction is issued, the party against which the injunction has been granted has a tendency to extend the prosecution and the verdict. Approximately 9% of the cases were finally resolved because the stay order was issued more than 20 years ago. About 21% of the cases were resolved because the stay order was issued more than 10 years ago. There are currently over 3,000 cases pending in different courts spanning more than 10 years, according to records available till 2017.

4.1.2 PROCESS FLOW CHART OF PATENT PROSECUTION

Nation has its own set of rules for accepting patent applications, reviewing them, and then granting or denying patent grants. The time it takes for a patent to be awarded to your invention can vary depending on the process adopted by a patent office. The Indian Patent Office (IPO) also follows a set of guidelines, and this chapter outlines the following steps and deadlines involved in patenting invention in India (Figure 4.2):

- The process of filing a patent in India
  Give the tax-paying documents to the fee counter and the nonpaying documents to the nonpaying counter after the application has arrived at the patent office.

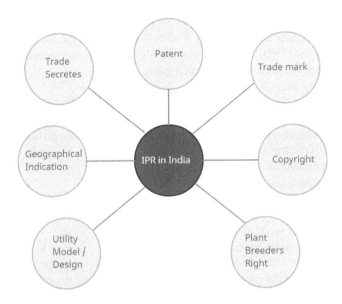

FIGURE 4.1 Categories of intellectual property.

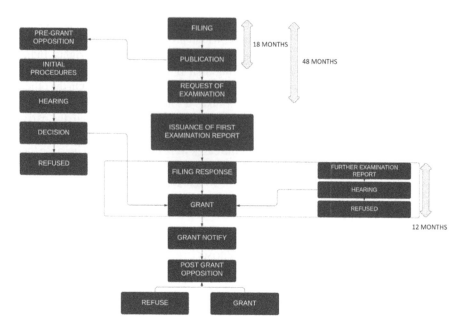

FIGURE 4.2 Process flowchart of patent prosecution in India.

- Employees enter pertinent information into units, seal papers, and produce Cash Book Receipts (CBRs). They fill in the CBR number, the year, the amount of the fee paid, the filing number, the patent number, and other relevant information.
- The nonfee counter team then enters pertinent information in the paper acceptance module and seals the records that have been issued.
- Send records for digitization from both counters to the Electronic Data Processing (EDP) section. And non-digital records are submitted to the appropriate sections on a regular basis.
- Preliminary analysis
    The office assigns a date and serial number to each application upon receipt, Patent implementations in e-wrapper format.
    Screening of applications based on the following criteria: (a) International Use; (b) Technical Area of Discovery
    - Defense/atomic energy relevance.
    - Make sure the abstract is correct or complete.
        Following that, they number the applications based on the year of filing, authority, and type of submission. The filing number is a six-digit continuous serial number that can be used by all IPOs.
- Examination of the application
    The patent office has certain guidelines for how a patent application passes through the patent office. Before submitting your submission, you should be aware of the scrutiny requirements. The following are some of the inspection criteria:

- Filing in appropriate jurisdiction: the selection of appropriate jurisdiction is one of the requirements used to scrutinize patent prosecution in India. If the information is incorrect, the submission will be rejected, and you will be notified of this.
- Proof of right: it must be submitted within six months of the application's filing date if it was not filed with the application. Alternatively, you may file it with a petition under Law 137/138.
- Paper completion and specification: the office verifies that the submission is complete with all necessary documentation and is in a correct format. It also verifies that the claimant, inventor(s), or agent has signed it properly (if any).

- Application publication

  From the filing date or priority date, it takes 18 months for the paper to be written (whichever is earlier). The office journal is used for publication. You may also order early publishing of your submission by completing Form 9 and paying the required fee. In this scenario, the publication occurs within one month of the submission date. After the publication of your patent claim, a third party will file a pre-grant opposition. They will challenge the IPO's patentability requirements, which include creativity, nonobviousness, nonpatentable subject matter, and nondisclosure.

- Examining your application

  This is the most important step in patent prosecution, whether in India or elsewhere. A submission for examination is the first step in the application examination process. This is accomplished by submitting a completed form-18 or form-18A (for startups) along with the review fee (electronic transmission).

  - Send the proposal within 48 months of the filing date or the priority date, whichever comes first.
  - If this does not occur, the submission would be deemed withdrawn.
  - No investigation takes place until an order for examination is published and filed.
  - When reviewing the submission, the patent officer searches at all of the patentability requirements. If the submission fails to meet those requirements, it will be removed from consideration.
  - The office controller generates an office intervention where there are small errors such as grammatical or format-related errors. Within a 3-month cycle, you must refresh the application by fixing the errors.
  - The patent office issues a minimum of two office acts; however, you may recommend that the application be examined further.
  - If your patent application complies with the patent office's laws, you will be granted a patent for a term of 20 years; if it does not, your patent application will be rejected.

- Payment of annuity

  To keep your patent enforceable, you must pay an annuity premium. It's often referred to as a renewal charge or a repair fee. Also, find out how much it costs to register a patent in India.

- Opposition to the grant

    A third party may file a post-grant opposition one year after the patent is granted. He or she will challenge whether the idea meets the patentability requirements, much as in pre-grant opposition. The judicial process could then result in the patent being revoked.

4.2 INTELLECTUAL PROPERTY RIGHTS

IPRs are civil rights that cover discoveries and/or discoveries made as a result of intellectual work in the fields of science, literature, industry, and arts. Copyrights, patents, trade secrets, and trademarks are the most popular IPRs.

IP includes ideas, innovations, industrial buildings, patents, music, literature, symbols, titles, labels, and other items produced by the human intellect. Most kinds of property rights are related to IP rights. They enable the proprietor to benefit completely from his or her product, which started as an idea and ultimately became a reality. They often provide him or her the right to prevent others from using, playing with, or interfering with his or her property without his or her permission. He or she has the legal right to sue them and demand that they stop compensating him or her for any losses incurred.

IPR has a long and illustrious history. IPR is believed to have started during the Renaissance in North Italy. Venice established a statute regulating patent rights in 1474, giving the owner an exclusive right to use the invention. The printing press with replaceable/movable wooden or metal letters was developed by Johannes Gutenberg in 1440 AD, giving rise to the patent. In the late 1800s, a number of nations were forced to pass laws regulating IP rights. Two international conventions have been signed: the Paris Convention for the Protection of Industrial Property (1883) and the Berne Convention for the Protection of Literary and Artistic Works (1886).

IPR security

Although IPR defense does not mean intellectual monopoly, it does enable the inventor, trademark owner, patent holder, and copyright holder to profit from his or her work, labor, and investment. The International Declaration of Human Rights outlines the freedom to benefit from the protection of moral and physical interests resulting from the right holder's work: literal or creative product.

IPRs are designed to encourage the creation of new works, such as technology, artwork, and inventions that may contribute to economic growth. Individual's incentives to continue creating goods that create new employment and emerging technologies are increased by IP rights, enabling our society to evolve and grow much more rapidly.

4.3 TYPES OF IPR

- a. Patents.
- b. Trademarks.
- c. Copyrights and related rights.
- d. Geographical indications (GIs).
- e. Industrial designs.

f. Trade secrets.

g. Layout design for integrated circuits.

h. Protection of new plant variety.

4.3.1 PATENTS

In general, a patent is a license given to an author on his or her product, enabling him or her to commercialize and exploit it in the industry over a fixed amount of time to the exclusion of others. Any fresh and beneficial thing is called "innovation."

i. material, device, procedure, or method of production;

ii. apparatus, machine, or other article;

iii. Any new and beneficial improvement of any of them, as well as a claimed innovation, is material produced through manufacturing, according to Section 2(1) (j) of the Patents Act, 1970.

In the Patents Act of 1970, the term "invention" refers to both a new product and a new method. As a result, in India, a patent may be submitted for a "Product" as well as a "Process" that is unique, involves an innovative phase, and is capable of being used in industry

The technology would not be deemed new if it was already revealed a written or spoken explanation to the general people in India or anyplace else in the globe, usage, or other means before the patent application was filed. Prior experience will also include material found in newsletters, scientific journals, and books. A patent cannot be awarded if the technology is still part of the state of the art. Displaying goods at conventions, trade shows, and other events when discussing how they operate are examples of such disclosures, as are related disclosures in an article or publication.

It's worth mentioning that any invention that fits into one of the criteria below isn't patentable: (a) frivolous, (b) apparent, (c) in violation of well-established natural laws, (d) in violation of law, (e) in violation of morals, (f) injurious to public health I a simple discovery of some new property or new use for a known substance, method, machine, or apparatus, (g) a simple discovery of a scientific principle, (h) the formulation of an abstract theory, (i) a simple discovery of some new property or new use for a known substance, method, machine, or apparatus, (j) a material created by a simple mixing that only produces the inventor, or the assignee or legal guardian of the inventor, may apply for a patent. The patent in India is valid for 20 years. The patent is renewed every year since it is applied.

Using technology or inventing something new

When using some technology or innovation, the startup can double-check that it would not infringe on the patentee's rights. If the startup wants to use a proprietary invention or technology, it must first obtain permission from the patentee.

- Patent rights compliance

    It is important to remember that patent infringement litigation in India can only be started after the patent has been granted, but it can contain an argument that dates back to the date on which the patent application was published. Infringement of a patent occurs when a copyrighted invention is made, imported, used, offered for sale, or sold without permission within India. Only a tort suit may be filed before a court of law under the (Indian) Patents Act, 1970. A court may award permanent and temporary injunctions, fines or account of earnings, delivery of infringing products for destruction, and expense of legal action in a suit for patent infringement, similar to trademarks.

• Patents are needed

Individuals profit from patents because they are acknowledged for their creativity and are compensated financially for commercially successful innovations. These rewards stimulate creativity, ensuring that human life quality continues to improve.

• Importance of patents in daily life

In fact, patents have pervaded every aspect of human existence, from electric lighting (Edison and Swan patents) and plastic (Baekeland patents) to ballpoint pens (Biro patents) and microprocessors (Intel patents) (patents held by Intel, for example). Both patent owners must willingly disclose information regarding their innovation in order to add to the world's total body of science knowledge in return for patent privileges. A body of public awareness encourages others to be more creative and innovative. Patents thus offer not only rights to their owners but also useful knowledge and encouragement to future generations of scholars and inventors.

• What is the procedure for obtaining a patent?

The first step in acquiring a patent is to file a patent application. The invention's title, as well as a description of its technical field, must be included in the patent application, as well as the history and a description of the invention in plain language and sufficient detail for an individual with a working knowledge of the field who can use or reproduce the innovation. To best explain the invention, those accounts are typically supplemented by graphic materials such as sketches, plans, or diagrams. The filing also has a number of "claims," or pieces of evidence, that define the scope of the patent's defense.

• Inventive concepts that could be covered

To be covered by a patent, an invention must, in general, meet the following requirements. It must be useful; it must have a component of novelty, i.e., any new feature that is not recognized in the body of established information in its technological area. The term "prior art" refers to the body of current information. The innovation must demonstrate an innovative move that a person with average technological expertise would not be able to deduce. Finally, the subject matter must be deemed "patentable" in the eyes of the statute. Science models, statistical techniques, plant or animal types, natural drug discoveries, commercial procedures, or medicinal procedure methods (as opposed to medical products) are usually not patentable in many countries.

• Patents are granted by

A patent is issued by either a national or a regional patent office, such as the European Patent Office or the African Intellectual Property Organization, which operates on behalf of other countries. A patent application is filed in one or more countries, and each nation determines whether or not to provide patent protection inside its boundaries. The Patent Cooperation Treaty is governed by the World Intellectual Property Organization (WIPO), and it allows for the submission of a single international patent application that has the same effect as national applications submitted in the specified nations. A defense claimant should file a single application to apply for protection from as many signatory states as they require.

4.3.2 TRADEMARKS

The Trade Marks Act of 1999 (TM Act), for example, allows for the registration of trademarks, the filing of multiclass applications, the renewal of a trademark registration for a period of ten years, and the acknowledgment of the concept of well-known marks, among other things. It's worth noting that the letter "R" in a circle, i.e., ®, can only be used for a trademark after it's been registered under the TM Act.

Trademarks are any words, phrases, emblems, slogans, product packaging, or architecture that identifies goods or services from a certain source. As defined in Section 2(zb) of the TM Act, a "trademark" is a mark capable of being graphically represented and capable of distinguishing one person's products or services from those of another, and may include the form of items, their packaging, and color combinations. The TM Act's trademark concept is broad enough to include nontraditional markings such as color marks, sound marks, and so on. According to Section 2 (m) of the TM Act, a "mark" is defined as a device, brand, heading, label, ticket, name, signature, phrase, letter, number, form of goods, packaging, or a combination of colors or any combination thereof.

As a result, any symbol used by a startup in the trade or industry in any way for the purpose of separating itself from others may be considered a trademark. It's worth noting that the Indian judiciary has taken a constructive approach to trademark rights, extending trademark protection to domain names in seminal cases such as *Tata Sons Ltd v Manu Kosuri & Ors* [90 (2001) DLT 659] and *Yahoo Inc. v Akash Arora* [1999 PTC 201].

- Consider these factors when choosing a trademark

    Any startup should exercise caution when choosing its trade name, labels, colors, product packaging, domain names, and any other trademarks. Before registering a trademark, you must do a thorough research. The following five sections can be used to classify trademarks:

    a. Generic
    b. Descriptive
    c. Suggestive
    d. Arbitrary
    e. Invented/Coined

        Generic labels refer to using the product's branding on the product, such as "Salt" for salt.

        Descriptive marks are those that describe a product's attribute, such as the symbol "Equal" for fairness creams.

Suggestive markings are those that suggest a product's feature, such as "Habitat" for home furnishings.

Arbitrary tags, such as "Blackberry" for mobile, remain in traditional lexicon but have no logical connection to the products or services for which they are used.

The term "invented/coined logo" refers to the creation of a new word that has no dictionary definition, such as "Adidas." Invented or arbitrary markings are the most powerful and therefore the simplest to defend.

The weakest markings are those that are abstract or suggestive in nature and are difficult to defend. Generic brands are the weakest, since they will never be used as trademarks.

For trademark registration in India, the NICE Classification of Goods and Services is used. The NICE Classification divides items into 45 different categories (classes 1–34 include goods and classes 35–45 include services). The NICE Classification is accepted in the majority of nations and streamlines the trademark registration process globally. Any startup that wants to trademark a product or service must choose the right class from a list of 45 options.

When choosing a mark, the startup should bear in mind and check whether the trademark is being used by anybody else in India or overseas, particularly if it is well-known. It's worth noting that India understands the idea of a "well-known trademark" and the definition of "trans-border reputation."

Google, Tata, Yahoo, Pepsi, Reliance, among other well-known trademarks are examples. Furthermore, trademarks such as Apple, Gillette, Whirlpool, and Volvo have been granted rights in India under the concept of "Trans-border Reputation," even though they have no physical presence in the country.

• What are the different types of trademarks that can be registered?

The options are practically endless. Trademarks may consist of a single word, letter, or numeral, or a mixture of words, letters, and numerals. Drawings, patterns, three-dimensional signals like the form and wrapping of products, auditory signs such as music or voice effects, fragrances, or colors used as identifying features are all examples. Several other types of logos occur in addition to trademarks that designate the industrial origins of products or services. An association owns collective labels, which its members use to distinguish themselves with a certain standard of quality and other criteria defined by the association. Accountants, builders, and architects, for example, are represented by those organizations. Certification marks are awarded for meeting defined criteria, but they are not tied to any particular affiliation. They should be sent to someone who can certify that the goods in question follow those criteria. These commonly recognized certifications include the globally agreed "ISO 9000" consistency standards.

- How to register a trademark

  To begin, an application for trademark registration must be sent to the relevant national or regional trademark office. A simple copy of the sign filed for approval, including any colors, shapes, or three-dimensional elements, must be included in the filing. A selection of products or services on which the sign will refer must also be included in the document. To be protected as a trademark or other form of symbol, the sign must meet certain criteria. Consumers must be able to differentiate the brand from other trademarks representing other brands, as well as associate it with a specific product. It must not confuse or manipulate consumers, nor must it breach good order or morals. Finally, the protection sought cannot be identical or equivalent to those previously given to any trademark owner. This may be decided by the national office doing a search and inspection, or by the opposition of third parties claiming equal or equivalent rights.

- Trademarks protection

  Trademarks are registered and protected in almost every country on the planet. Each national or regional office maintains a Register of Trademarks that contains full application documents for all registrations and renewals, allowing third-party examination, scan, and future opposition simpler. However, the results of such a registration are restricted to the country (or countries, in the case of a geographical registration) in question.

- Trademark rights compliance

  Trademarks may be covered by both statute legislation, such as the common law and TM Act, such as the passing off solution. If anyone uses a similar logo for similar or related products or services, or whether they use a well-known trademark, the other party may sue for infringement of IP rights, regardless of whether the trademark is licensed or not. In India, trademark registration is not a requirement for pursuing a civil or criminal suit for trademark infringement. Under trademark rules, the prior adoption and usage of the trademark is important. In a suit for violation or passing off, a court may typically issue permanent and temporary injunctions, damages or a profit account, delivery of the offending product for destruction, and expense of the legal proceedings. It is important to remember that trademark infringement is a felony offence for which criminal charges can be brought against the infringers.

4.3.3 COPYRIGHTS AND RELATED RIGHTS

The legal right of an author, artist, or creator to commercially exploit his or her original work that has been expressed in a physical manner, as well as the right to prevent such material from being reproduced or reused without his or her consent, is referred to as copyright.

A picture, a sculpture, a sketch (including a diagram, a plot, a chart, or a plan), an engraving, an image, a work of design or creative craftsmanship, a theatrical work (recitation, choreographic work), and literary work (including computer programs, charts, compilations, and computations) are all examples of works on which copyright is still retained under the Copyright Act of 1957.

- The duration of copyright in the case of original literary, dramatic, musical, and artistic works is the author's or artist's lifetime plus 60 years from the year of the author's or artist's death; in the case of cinematograph films, sound recordings, posthumous publications, anonymous and pseudonymous publications, works of government, and works of international organizations, the duration of copyright is the author's or artist's lifetime plus 60 years from the year of the author's or artist's death.
- The Copyright Act of 1957 brought India's copyright law in step with advances in the digital technology industry, whether it be satellite transmission, computer applications, or new technology, in order to meet the worldwide need for harmonization.

Copyright registration

In India, copyright registration is not required because it is regarded as merely a record of a fact. The registration does not grant or establish any additional rights, and it is not used to take legal action against infringers. The Indian courts have supported this viewpoint in a series of rulings. Although copyright registration is not required in India and is protected by the International Copyright Order, 1999, it is recommended that copyright be registered because in courts and by police forces, the copyright registry certificate is known as "proof of ownership" and is worked upon smoothly.

What are some of the advantages of copyright and related rights?

Under copyright law, the authors of works protected by copyright, as well as their descendants and successors (collectively referred to as "rights holders"), have some fundamental rights. They have the exclusive right to use or allow others to use the job under such conditions. The owner(s) of a work's rights can prevent or permit its replication in any medium, printing and sound recording; public exhibition and contact with the public; broadcasting; translation into other languages; and adaptation, such as a book into a screenplay. Linked rights grant similar rights to fixation (recording) and reproduction, among other things. Many types of works, such as publications, sound recordings, and films, which are protected by copyright for effective reproduction, and associated assets, necessitate widespread production, correspondence, and financial investment; as a consequence, authors often sell the rights to their creations to corporations ideally suited to produce and distribute them in return for payments in the form of a royalty (remuneration calculated as a proportion of the work's income).

- Advantages of copyright and associated property protection
    The defense of copyright and associated rights is critical in promoting human imagination and innovation. Giving writers, artists, and producers benefits in the form of praise and equal monetary compensation raises their production and, in many cases, improves the outcomes. Furthermore, by ensuring the presence and enforceability of rights, businesses and organizations can more effectively engage in the production, growth, and global

distribution of works; this, in turn, aids in increasing access to, and improving the enjoyment of, culture, information, and entertainment across the world, while also promoting economic and social development.

- How have copyright and associated protections held pace with technological advancements?

  With the spectacular success of technical advances over the last few decades, the area of copyright and associated rights has grown tremendously, bringing new ways of disseminating works through such means of worldwide communication as satellite radio, compact discs, and DVDs. The distribution of works over the Internet is just the most recent creation, raising new concerns in this global medium with regard to copyright and related protection.

- What laws govern copyright and associated rights?

  Without the requirement for registration or other formalities, copyright and associated rights protection is gained automatically. Many nations, on the other hand, have a national scheme of optional registry and deposit of works, which can help with issues such as possession or production conflicts, funding transactions, contracts, allocations, and transfers of property. Many writers and performers lack the opportunity or resources to seek copyright and related rights by legal and administrative means, particularly given the growing global use of literary, artistic, and performance rights. As a result, the creation and enhancement of collaborative management organizations, or "societies," is a rising and significant movement in many countries. Members of these societies can benefit from the organization's administrative and legal expertise, as well as continuity in obtaining, managing, and disbursing royalties received from national and foreign usage of a member's work or performance, for example. Producers of sound recordings and media companies have such privileges that are often administered jointly. Anyone who uses another person's original work without first seeking permission from the author infringes on the owner's copyright. In India, copyright law provides not only civil remedies such as damages or accounts of profits, permanent injunctions, delivery of infringing material for destruction, and legal costs, but it also makes copyright infringement a cognizable offence punishable by imprisonment for not less than six months but not more than ten years.

  Under the Copyright Act, there are provisions for increased fines and penalties for the second and third offences. The (Indian) Copyright Act, 1957, empowers the police to file a lawsuit (First Information Report, or FIR) and act individually to apprehend the accused, check the accuser's property, and recover the infringing content without the need for court interference.

4.3.4 GEOGRAPHICAL INDICATIONS

GIs are labels that are used on products that have a certain geographical origin and that have characteristics or a reputation that are unique to that location. Agricultural products usually have characteristics that are determined by particular local conditions, such as temperature and soil, and are derived from their place of origin. They may also call attention to product features that are linked to human influences in the product's origin, such as engineering expertise and customs.

A GI identifies a particular location or area of production that specifies the distinctive characteristics of the commodity produced there. It is critical that the product's qualities and prestige are derived from that location. A village or district, a city, or a country may be the place of origin.

Since it is an exclusive privilege granted to a certain group, the privileges of its registration are enjoyed by every community member. Recently, GIs for items such as Chanderi sarees, Kullu shawls, and Wet Grinders were reported. Given the wide range of traditional goods available across the world, registration under the GI would be critical to the future growth of the tribes, societies, and professional artisans involved in their development.

- What is the need for geographical signs to be protected?

    Consumers recognize geographic signs as indicators of product origin and quality. Many of them have built desirable reputations that, if not properly safeguarded, may be misrepresented by unscrupulous commercial operators. Unauthorized use of GIs, such as "Darjeeling" for tea that was not cultivated in Darjeeling's tea gardens, is harmful to consumers and legal farmers. The former are duped into thinking they are purchasing a real commodity with unique attributes and features, when in reality they are receiving a worthless knockoff. The latter suffers losses as a result of lucrative companies being stripped away from them, as well as a tarnished name for their goods.

- What's the difference between a trademark and a GI?

    A trademark is a sign that differentiates the products and services of a business from those of rivals. It gives the owner the right to forbid them from using the trademark. Customers can tell whether a product was produced in a certain area and if it has any characteristics that are unique to that location. It may be used by any manufacturer who makes their goods in a geographically defined area and whose products have similar characteristics.

- What safeguards are in place for GIs?

    Geographical signs are governed by national legislation and a variety of concepts, including unfair competition laws, consumer rights laws, credential mark protection laws, as well as special legislation protecting GIs or origin appellations unauthorized gatherings are, in essence, are prohibited from using geographical signs if they are likely to confuse the public about the product's true origin. Judicial injunctions against unlawful usage are available, as well as the payment of penalties and fees, including, in severe circumstances, incarceration.

- How are geographical signs safeguarded on a global scale?

    The Paris Convention for the Defense of Industrial Property of 1883 and the Lisbon Agreement for the Protection of Appellations of Origin and Their International Registration, both of which are governed by WIPO, include geographical signs.

- What does it mean to have a "generic" GI?

    When a place's name is used to designate a certain type of product rather than as an indicator of the product's place of origin, the word loses its regional significance. For example, "Dijon Mustard" is a type of mustard that originated several years ago in the French town of Dijon, but has since come to refer to a specific type of mustard produced in a variety of locations. As a result, "Dijon mustard" has become a common term that refers to a category of substance rather than a specific location.

    The signs must come under the limits of Section 2(1) of the GI Act, 1999, in order to be registered. As a result, it must also meet the requirements of Section 9, which prohibits the registration of a GI:

- Whose use would be likely to deceive or cause confusion; or
- Whose use would be contrary to any law in force at the time; or
- Can include or incorporate indecent or scandalous material; or
- Which comprises or contains any matter likely to harm the time?

    "Generic names of indications" in reference to products that have lost their original significance and have become the general name of such goods and function as a designation for an indicator of the sort, value, form, or other property or feature of the goods, even though they refer to the location of the area where the goods were originally produced or made.

    "All considerations, including the current situation in the country or location where the name originates and the area of consumption of the products, shall be taken into consideration in deciding whether the name has become generic."

Flowchart of the registration process

The Geographical Indications of Goods (Registration and Protection) Act 1999 was passed by Parliament in December 1999 (Figure 4.3). This Act aims to make it easier to register and secure GIs for products in India. The Registrar of Geographical Indications, who is also the Controller General of Patents, Designs, and Trade Marks, is in charge of enforcing this law. Chennai is home to the Geographical Indications Registry.

There are two sections of the Registrar of Geographical Indication. Part 'A' contains information about registered geographical signs, while Part 'B' contains information about registered approved users. The registration method is identical to that seen above for registering a GI and a registered person.

4.3.5 INDUSTRIAL DESIGNS

A novel or initial design that is given to the owner of a validly licensed design is referred to as a design privilege. Industrial designs are pieces of art that result in the ornamental or structured appearance of a commodity, whereas a design right is a novel or initial design awarded to the owner of a validly registered design. Industrial designs are considered to be a type of IP. Minimum levels of safety for industrial designs have been established under the TRIPS Agreement. India, as a developing

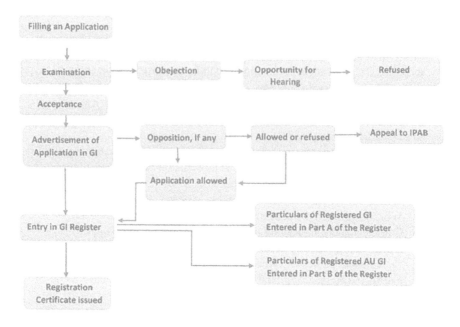

FIGURE 4.3 Registration process of geographical indications.

country, has also revised its national laws to include these minimum requirements. Design law's main goal is to encourage and preserve the aesthetic component of industrial development. It's also meant to encourage industry-related innovation. The New Designs Act, 2000, is India's current industrial design law, and it will serve its role well in the face of rapid technological progress and foreign innovations. India has now reached a mature position in the area of industrial designs, and in light of the economy's globalization, the current law has been updated to reflect the changing technological and commercial landscape and to adhere to international architecture administration patterns. This replacement Act further seeks to introduce a more comprehensive definition of architecture in accordance with international standards, as well as to address the expansion of design-related practices in various fields.

- Why should industrial designs be protected?

    Industrial designs increase a product's economic worth and marketability by making it more desired and attractive. When an industrial design is protected, the manufacturer—the person or business that registered the design—is given an exclusive right against unlawful copying or imitation by other parties. This ensures a fair return on investment. By fostering equal markets and honest trading practices, stimulating innovation, and promoting more aesthetically appealing goods, a successful scheme of defense benefits consumers and the general public. Protecting industrial designs encourages innovation in the industrial and engineering industries, as well as in conventional arts and crafts, which continues to drive economic growth. They aid in the growth of industrial operations and the sale of domestic goods.

Industrial designs can be developed and protected in a relatively simple and low-cost manner. Small- and medium-sized businesses, as well as independent artists and craftsmen, in both developed and developing nations have fair access to them.

- How are manufacturing designs safeguarded?

In certain nations, an industrial design must be licensed in order to be protected by industrial design legislation. To be registrable, a design must be "new" or "original" in general. Different nations have different meanings for these terms, as well as different registration procedures. In general, "new" means that no concept that is equivalent or somewhat similar to it has ever been used before. A registry certificate is issued after a concept has been registered. Following that, the duration of immunity is usually 5 years, with the option of further expiration periods of up to 15 years in most cases.

The sector is responsible for the high quality of life enjoyed by people in the developing world today. All were a one-of-a-kind handmade objects before production. Then came mass manufacturing, which permitted a community of people to produce vast numbers of similar items, from clothes to automobiles, with the help of machinery. Of course, without engineers to streamline and refine the mass manufacturing process, none of this would have been possible—industrial designers became the driving force behind the Industrial Revolution. These pioneers were more than engineers; they were also artists who figured out how to create a large number of products cheaply while also making them attractive enough for customers to want to purchase them. Industrial designers take a functional product that fulfills a need and improve it by making it more useful or attractive. They are the ones who design modern vehicle models or make improvements to the ergonomics of a chair or a computer keyboard. They keep all of our everyday conveniences unique and exciting, making them ever more so as time passes.

4.3.6 TRADE SECRETS

A trade secret is a method, process, mechanism, design, tool, pattern, commercial system, or collection of information that enables a business to obtain a competitive or customer-related economic advantage that is not generally known or readily ascertainable by anyone. There is an almost infinite variety of trade secrets. As a consequence, any knowledge that is not generally known in relevant business circles or by the broader public is classed as a trade secret. Furthermore, the substance should be difficult to obtain.

- Provides a form of financial advantage to the holders. This advantage must be derived directly from the fact that it is not well recognized, rather than simply from the information's meaning. Since it is a password, it must have market appeal. Commercial appeal includes both future and real value. The lawful owners of the knowledge must have taken appropriate precautions to keep it confidential. What constitutes reasonableness varies depending on the circumstances.

A trade secret is a trade secret for as long as the knowledge is kept secret. However, as material becomes widely available, it is no longer sufficiently secured, or has no market significance; it can no longer be considered a trade secret. One of the most important assets in a company's IP portfolio is trade secrets. Trade secrets are treated in the same way as other forms of IP such as patents and trademarks. Trade secrets, according to others, are the crown jewels of every IP portfolio. A trade secret is any proprietary commercial knowledge that provides a strategic advantage to a company. Unauthorized possession of any material by people other than the holder is considered an unethical activity and a trade secret breach. Depending on the state, trade secret privacy is either used in the broader definition of unfair market protection or is focused on special provisions or case law relating to the protection of sensitive information. A trade secret is a sensitive piece of knowledge that is kept classified and provides a strategic edge to an organization. Misappropriation of trade secrets is becoming more common in European industries. The European Commission is aiming to harmonize the current disparate national rules on trade secret security so that businesses can exploit and exchange their trade secrets with privileged business partners across the EU, resulting in increased productivity and job creation (Figure 4.4).

The Uniform Trade Secrets Act, or UTSA, is now a single trade secret statute in the United States. Surprisingly, state legislation governs the majority of trade secret cases. Except for Massachusetts and New York, every state has adopted some kind of UTSA. UTSA's vocabulary, as well as state laws that have adapted it, is somewhat similar to TRIPS' (Trade-Related Aspects of Intellectual Property Rights) language. TRIPS is a World Trade Organization (WTO)-managed international agreement that sets uniform standards for some kinds of IP regulation that apply to nationals of other WTO members.

Trade secrets have many advantages.

IP can take several forms. The following are some of the benefits of trade secret protection:

- There are no licensing fees for trade secrets. Obviously, the administrative, technological, and/or legal obstacles that the corporation erects to protect its trade secrets may incur costs.

FIGURE 4.4 Trade secrets.

- Unlike a trademark, which becomes public property, trade secret security does not entail publication or certification.
- Trade secrets are not restricted in length, unlike a patent, which is only valid for 20 years; and
- Trade secrets take hold immediately, unlike a patent, which can take a few years to be issued.

4.3.7 LAYOUT DESIGN FOR INTEGRATED CIRCUITS

India passed the Semiconductor Integrated Circuits Layout-Designs Act, 2000, to protect integrated circuit layout designs in accordance with the TRIPS Agreement. A configuration of transistors and other circuitry components represented in some form in a semiconductor integrated circuit, including lead wires connecting specific elements, is defined in the Act. A Semiconductor Integrated Circuit is defined as a product with transistors and other circuitry elements that are inseparably shaped on a semiconductor material, an insulating material, or inside the semiconductor material and configured to perform an electronic circuitry function, as defined by the Semiconductor for Integrated Circuits Layout-Designs Act of 2000.

4.3.7.1   Layout-Design Register Ability

It is not possible to register a Layout Design if it is not an initial, if it has been commercially used in India or in a convention region, or if it is not intrinsically unique or identifiable from any other documented Layout Design. It is necessary to register Layout Design in order to claim rights to it.

4.3.7.2   Security Timeframe

A Layout- Design's registration is only valid for 10 years from the date of filing an application for registration or commercial exploitation in any jurisdiction, whichever comes first.

4.3.7.3   Layout-Design Infringement

In India, a violation of a Layout-Design is considered a criminal offence in addition to the civil remedies applicable. Under the (Indian) Semiconductor for Integrated Circuits Layout-Designs Act, 2000, infringement of a licensed layout design is punishable by up to three years in prison or a fine of Rs. 50,000 (approximately US$800) up to a limit of Rs. 1,000,000 (approximately US$16,000), or both.

4.3.8 PROJECTION OF NEW PLANT VARIETY

The aim of this act is to reward farmers for their position as cultivators and conservers of the country's agro biodiversity, as well as traditional, agricultural, and tribal communities' contributions to the country's agro biodiversity, and to encourage investment in R&D for the production of new plant varieties to help the seed industry expand.

Farmers' Rights and Plant Variety Protection Act 2001 was passed in India to safeguard new plant varieties, and it went into effect on October 30, 2005. Rice, wheat, maize, sorghum, pearl millet, chickpea, green gram, black gram, lentil, kidney bean, and other crop species were initially described for registration. India has selected a sui generis system over patents to ensure the safety of novel plant types. The administrative ministry in charge of registration and other matters is the Department of Agriculture and Cooperation.

4.4   ENFORCING IP RIGHTS IN THE UNITED STATES AND INDIA

In the United States, IP laws are enforced. To the rest of the country, the Office of International Intellectual Property Enforcement (IPE) reflects America's brilliance. IP is the lifeblood of our culture, reflecting America's creativity. The Office of Intellectual Property Regulation (IPE) aims to safeguard and enforce IP rights (IPR) throughout the globe. The IPE team works closely with US ambassadors and negotiators throughout the globe to ensure that the interests of American rights holders are heard, and to highlight the vital role that IPR security plays in fostering global innovation and economic growth.

IPE advocates for robust IPR systems in order to

- prevent users from being exposed to imitation and pirated goods;
- guarantee that the interests of American IPR holders are secured abroad; and
- foster IP security and compliance as a critical component of economic growth.

The development and regulation of IPR allows the United States to be one of the world's most creative nations. Microchips, the Internet, the telephone, the biotechnology revolution, sending humans to the Moon, and saving millions of lives by revolutionary drugs were all invented by American scientists, architects, and entrepreneurs. Writers, actors, film directors, and gamers from the United States have captured the imaginations of people all over the world.

- A robust framework of IPR ensures that inventors, industrial designers, and creative artists' designs are secured. They will be compensated for their work and continue to participate in new technologies if they want to. As a result, they are driven to share their knowledge and creativity with others, enabling others to come up with novel and innovative ideas.
- There is a lot of innovation all over the world. IP may be created by someone who sees a problem and wants a solution. It is important to promote both domestic and foreign creativity. IP preserves the ideas and incentives of American innovators while also stimulating the imagination of inventors and designers around the world by promoting the introduction of robust IPR programs and regulation around the world.

India's IP laws are enforced

India has a well-established regulatory, institutional, and judicial system for protecting IPRs, but regulation remains a challenge. The sluggish judicial system in

India, which involves lengthy and time-consuming trial procedures, has long been a source of concern; in recent years, however, Indian courts have shown vitality and enthusiasm in successfully defending IPR. It has been found that IPR can be successfully secured with the assistance of law enforcement agencies by implementing the appropriate policies and strategies. It is important to consider the Indian judicial system and its psychology before engaging in any IPR-related litigation. It has been noted that while Indian courts are quick to issue equitable relief such as injunctions, they are also hesitant to pay punitive damages.

4.5   SELF-HELP CONSIDERATIONS

You can do a few things to make it more difficult for infringers to clone the product in general. When recruiting new workers, consider the quality of the product and how easy it will be for anybody to duplicate it without seeing the original designs and have appropriate IP-related clauses in employment contracts. Make sure your employees are educated on IP rights and security; have solid physical protection and destruction methods in place for documents, drawings, tooling, samples, machinery, and so on; make sure there are no "leakages" of packaging that counterfeiters could use to pass off fake products; check production over-runs to ensure that a genuine product is not being sold as something else.

4.6   KEY ISSUES RELATED TO IP

Trade is becoming heavily reliant on ideas and information. Many items that were once considered low-tech goods or services today have a higher proportion of value attributed to innovation and design. Films, music albums, books, electronic applications, online services, clothes, food, plants, biotechnology devices, and a variety of other items are purchased and sold for the content, ingenuity, and identification they contain, not for the plastic, metal, fabric, paper, or other material used to create them.

4.6.1   PLAGIARISM

Plagiarism is where someone steals someone else's thoughts or vocabulary and claims them as their own. Cutting and pasting passages into term papers and other materials without adequate attribution or quotation marks has become simple thanks to the proliferation of multimedia content and the expansion of the Web. While several websites state that their programs can only be used for testing, many consumers ignore these alerts. As a result, plagiarism has become a problem at all stages of education, from primary school to graduate school. Plagiarism exists outside of academia as well. It has been accused of famous literary writers, playwrights, singers, journalists, and even software developers. Even though there are codes of ethics in force that simply describe plagiarism and recommend punishments, many students actually have little understanding what plagiarism is, amid repercussions varying from no credit on a paper to expulsion. Some students claim that all electronic materials are free to use, and others intentionally plagiarize whether they are under pressure to get a good grade or because they are too tired or pressed for time to do original work. Plagiarism by students enrolled in Coursera's free online courses has been prevalent.

Some of the steps that schools should take to combat student plagiarism are mentioned below:

- Assist students in understanding what constitutes plagiarism and the importance of properly referencing sources.
- Demonstrate to pupils how to record Web pages and resources from online libraries.
- Make it obvious to pupils that academics are aware of paper mills on the Internet.
- Break down large writing assignments into parts over the course of the semester to reduce the possibility of students being pressed for time and having to plagiarize to meet the deadline.
- Assist instructors in educating pupils about plagiarism detection software and making sure they understand how to utilize it.
- Develop a robust antiplagiarism network around identification tools and facilities.

4.6.2   REVERSE ENGINEERING

This section delves into the history of reverse X engineering in the sense of invention under trade secret regulation, as well as in the context of digital products under copyright, patent, and a hybrid regime. This section also looks at how reverse engineering can be used in appropriation art forms such as visual collage and digital audio sampling for aural music collage. However, it should be noted that developing a unified definition and implementing reverse engineering of electronic devices and productions is difficult. Patents, copyrights, and trademarks allow people to profit financially or get notoriety from what they innovate or produce. The IP system strives to establish an environment where creativity and innovation may flourish by finding the correct balance between inventors' interests and the public interest. DMCA have distinct approaches to reverse enginnering but also it includes referring to the same electronic products. As a result, reconciling contradictions and the crippling impact of a schizophrenic IP scheme on the whole is more essential than ever.

Reverse engineering is the method of examining a commercially accessible commodity explicitly to ascertain its total structure, component components, technology, and all other features that enable it to work in a legitimate transition. The reverse engineer examines and determines how the component was made by dismantling it. As a result of reverse engineering an existing device, a rival is able to discover production and design secrets to use in his or her own inventions. In trade secret law, reverse engineering is a well-established right that is regarded as a legal and equitable way to compete by obtaining trade secret knowledge about an invention.

Reverse engineering, according to the Supreme Court, is "an important aspect of invention." Under the equal use and misuse doctrines, copyright law has accepted reverse engineering for computer systems. In standard patent law, though, there is no such thing as a reverse engineering right. Reverse engineering is the practice of dismantling something in order to learn more about it, create a replica, or enhance it. Reverse engineering was first used to reverse engineer computer hardware, but it is now widely used to reverse engineer applications as well. Analyzing software to construct a new interpretation of the device in a particular way or at a higher level of abstraction is known as reverse engineering. Reverse engineering always starts with the extraction of design stage information from software code. The specifications of an information system at the design level are more conceptual and less specified than the computer code of the same system. One popular use of reverse engineering in computing is converting an application that ran on one vendor's database to run on another's (for example, from Access to Oracle). For application creation, database management systems have their own programming language. As a result, companies that wish to switch database vendors must rewrite current programs using the database programming language of the new vendor. The cost and time taken for this redevelopment could discourage a company from switching vendors, denying it the potential benefits of upgrading to a better database technology. A developer may use the code of the new database programming language to restore the design of an information system implementation through reverse engineering. The template will then be used to generate code (forward engineer) in the modern database programming language using code-generation tools. The time and expense of migrating the organization's databases to the modern database management system was significantly reduced thanks to this reverse-engineering and code-generation process. No one disputes the freedom to use this method to transform in-house applications.

After all, the businesses that use such apps built and own them. Using this method on a bought software product created and approved by third parties, on the other hand, is a different story. Since the app owner does not currently hold the access to the software, most IT administrators will consider this behavior immoral. Compilers and decompilers are methods that are used in reverse engineering. A compiler converts statements written in a source language (such as Java, C, C++, or COBOL) into machine language (a series of binary 0s and 1s that the computer understands). When a software company sells software to a customer, it usually does it in machine language. Reverse-engineering compilers, also known as decompilers, can read computer code and generate source code. Reverse Engineering Compiler, for instance, is a decompiler that reads an executable machine-language file and transforms the text to a C-like representation. Decompilers and other reverse-engineering methods may be used to expose the source code of a competitor's software; it may then be used to create a new software that mimics or interacts with the original. As a consequence, reverse engineering may be used to access material that has been copyrighted or designated by another business as a trade secret. The use of reverse engineering to allow interoperability has been approved by the courts.

4.6.3 OPEN SOURCE CODE

Open source software is described as software that is subject to the conditions of its accompanying license. Many people confuse open source software with public domain software; nevertheless, the public domain does not apply to open source software. The creator of public domain software has willingly lost all intellectual privileges, including copyrights, to the public. No license (i.e., permission to use) is required for the use of public domain software because all proprietary rights have been relinquished. As a result, programmers are free to use public domain software as they see fit, subject to the copyright holder's restrictions.

In reality, some open source software licenses have very unfavorable implications, making their usage riskier than that of conventional commercial software. Most licenses, for example, do not contain any indemnity for third-party infringement claims, unlike commercial software licenses.

"Viral" or "copy-left" licenses, on the other hand, require "derivative works" or modifications of open-source programs to be released under the same open-source license. The GNU General Public License version 2 contains these infectious terms, and it is one of the most widely circulated open-source licenses. Because the source code must be provided with the device and the licensee has the ability to modify and redistribute the software without penalty, the ramifications of such licensing may be severe. As a consequence, a corporation could surrender exclusive code privileges, be required to share trade secrets, and also lose the exclusive ability to use its own underlying code, which must be openly revealed in source code.

Open-source software is defined as software with publicly available source codes that can be accessed, modified, and redistributed by anyone. Closed source software source codes, on the other hand, are either not disclosed to the public or are only disclosed under strict confidentiality conditions. Open-source software helps to offer a robust and useful resource for developers and technology firms because it is publicly distributed and actively reviewed by the public. From the Google Chrome interface and Netflix's online television channels, all of the technological tools we use are based on open-source technologies. Established technology firms such as Microsoft and IBM have already made significant investments in open-source developers, implying that many more open-source projects will be deployed in the future.

Since license terms can take several different forms and do not enable you to explicitly accept or approve, they can be quickly ignored when purchasing new software items. However, before using or integrating new products into your own source codes, you can carefully research these license terms. If you want to produce proprietary software while keeping the source code private, you can contact the owner to confirm whether a proprietary license can be granted in return for royalties.

4.6.4 COMPETITIVE INTELLIGENCE

Competitive intelligence (CI) is the process of discovering, analyzing, and gathering patent and market data in order to gain insight into the industry's progress, competitors' strengths, and weaknesses. It offers information on competitors' strategies, products, and next moves. CI also assists in identifying business risks and opportunities, as well as providing market data to aid in the development of a strategic strategy for potential growth.

IP is interested with emerging and creative inventions that necessitate multimillion-dollar investments in patent approvals, registrations, and attorney's fees. In IP, CI provides detailed details about similar products and rivals, allowing you to sort through data and make decisions about how to progress your invention.

Investing money to trademark an idea is a gamble, since you risk losing some decent research-oriented field that the CI could protect. CI is used in a landscape analysis to gather information on competitive challenges, eliminate surprises, identify leading competitors, and discover potential opportunities. For R&D and evolving products and services, CI offers important patent data to rivals.

Surprisingly, CI can also be used to find competitors. Another competitor's patent application and its claimant, as well as other patent applications in the same patent paper, could be known to the competitor. CI also refers to the ability to conduct searches in which it can include key characteristics of branded goods and help applicants appreciate the scope of producing or commercializing a new product. CI will also include countries where the product has been successfully promoted or regions where the product has yet to be successfully marketed.

To learn more about how applicants are raising their IP, CI can be classified based on a competitor or a product/domain:

- Competitor-based CI focuses on a competitor's technical development, mergers/acquisitions, and branded products.
- Based on a domain or product top competitors in the domain/product, various technology used by competitors, and potential R&D areas for future advances are all included in the CI.

To summarize, CI plays a critical role in the IPR domain, assisting with decisions such as retaining and growing market share for a technology applicant, technological evaluation, obtaining licenses, monitoring top rivals and their devices, addressing technological challenges, and gathering valuable information for potential developments.

4.6.5 TRADEMARK INFRINGEMENT

Trademark infringement is the illegal use of a trademark or service mark. This may happen with goods or services, and it can lead to disinformation, deception, or uncertainty about the true source of a product or service. If they claim their trademarks are being infringed upon, trademark owners will file a lawsuit. A court injunction will prohibit a criminal from using a trademark if misuse is proved, and the owner may be granted monetary compensation.

Trademark infringement types

When investigating trademark infringement, it's important to understand that there are two forms of infringement:

1. Indirect infringement

    Section 29 of the Act defines direct violation. There are a few requirements for a direct violation to occur, and these are as follows:
    - Use by an illegal person: This ensures that a trademark is only infringed upon when it is used by anyone who is not approved by the registered trademark holder. It is not considered infringement if the symbol is used with the permission of the registered trademark holder.

- Same or resembling each other: The unauthorized person's trademark must be almost identical or deceptively close to that of a patent registration. The expression "deceptively identical" simply means that the average user can be fooled by the markings and mistake them for one another. Since the key word here is "may," it just has to be proved that this is a probability, not that it will necessarily happen. It is sufficient to prove violation where there is a risk of misrecognition of the symbols.
- Trademark registration: The Act only protects trademarks that have been registered with India's trademark registry. The common law of handing off is used to resolve conflicts where an unregistered label is violated. That is a tort rule that is applied where an individual or group of people suffers harm or harm to their goodwill as a result of their actions.
- Type of product or service: Unauthorized use of the logo for the propagation of products or services that come into the same class as the registered trademark is considered trademark infringement.

2. Direct infringement

In contrast to actual infringement, there is no clear clause in the Act dealing with indirect infringement. This isn't to say that there's no risk of indirect violation. The common law theory underpins the concept and application of indirect violation. It leaves not only the primary infringer liable, but also anyone who aids or causes the direct perpetrator to infringe.

Indirect violation can be divided into two categories:

- Responsibility by a third party: According to Section 114 of the Act, if a corporation performs a violation of the Act, the whole corporation is responsible. As a result, anyone responsible for the business, not just the principal infringer, would be liable for indirect violation, with the exception of those who behaved in good conscience and without knowledge of the infringement.

The following are the elements of vicarious liability: Where the individual has jurisdiction over the principal infringer's activities

- When the individual is aware of the violation and actively participates in it
- If the individual stands to benefit financially from the infringement

The only time a corporation is immune from vicarious liability for infringement is if it behaved in good conscience and has no knowledge of the infringement.

- Contributory violation: contributory infringement is made up of just three elements:
  - When the person is aware of the violation
  - Where the person contributes materially to the direct violation
  - Where the individual persuades the main infringer to violate the law.

There is no loophole in the case of contributory violation unless the contributory infringer has no chance of acting in good conscience.

4.6.6 CYBER SQUATTING

Cyber squatting is the process of registering Internet domains that are identical or close to a third-party business name or trademark in order to benefit from the goodwill of the third-party brand or to resell them for a profit. Cyber squatters take advantage of the domain name registry system's first-come, first-served existence to file as domain names third parties' trademarks, corporate names, or the names of prominent persons, as well as combinations thereof. The aim of cyber squatting is to transfer the domain name back to the copyright owner, profit from the goodwill of a third-party brand, or redirect site traffic to meaningless commercial transactions.

4.6.6.1 Recognizing the Signs of Cyber Squatting

Check to see whether the domain name leads to a website as a general rule. If it redirects you to a page that says "this domain name for rent," "under renovation," or "can't locate server," the chances are you're dealing with a cyber squatter. The lack of a functioning website may mean that the domain name owner's sole motivation for purchasing the domain is to resell it to you at a higher price. Of course, the lack of a website does not necessarily imply that a cyber squatter is present. There may also be a more benign reason, such as the domain name owner's completely legal intentions to launch a website in the near future. You could have a case of cyber squatting if the domain leads to a working website that is mainly composed of advertising for goods or services relevant to your trademark. If your company is well-known for offering audio-visual services, and the website you visit is crammed with advertisements for other companies' audio-visual services, the chances are good that the platform is run by a cyber squatter who is profiting from your company's fame by selling Google ads to your rivals.

4.7 CAPABILITY MATURITY MODEL INTEGRATION

CMM's counterpart, Capability Maturity Model Integration (CMMI), is a more advanced model that integrates the individual's finest features, CMM disciplines such as Software CMM, Systems Engineering CMM, People CMM, and so on. Integration of these fields to satisfy the requirements is difficult since CMM is a reference model of matured approaches in a single area. This is why CMMI is used since it allows for the convergence of several disciplines as required (Figure 4.5).

CMMI's goals are as follows:

1. Meet the wishes and desires of customers.
2. Create value for consumers and stockholders.
3. Accelerate the market's growth.
4. Improve product and service efficiency.
5. Improve industry credibility.

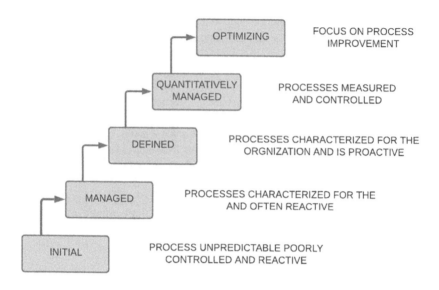

FIGURE 4.5   Characteristics of maturity levels.

Staged and continuous CMMI representation:
A company may use a representation to achieve a different set of improvement goals. CMMI can be represented in two ways:

- Staged representation: The enhancement route is specified by a predefined collection of process areas.
  - Offers a progression of changes, with each component serving as a basis for the next.
  - The sophistication standard determines whether or not a course is changed.
  - The maturity level of an organization's processes is defined by its maturity level.
  - A staged CMMI representation allows it possible to equate different individuals of different maturity levels.
- Continuous representation: This feature helps you to choose unique process areas.
  - Employs skill levels to monitor the progress of a single process sector.
  - Comparing various entities on a process-area-by-process-area basis is possible with continuous CMMI representation.
  - Enables companies to identify systems that need to be improved.
  - This diagram encourages organizations to choose the order in which they want to change different procedures, allowing them to achieve their goals and avoid risks.

Maturity levels in the CMMI model
There are five maturity stages defined in CMMI with staged representation:

1. Initial maturity stage
   - The systems aren't well-managed or supervised.
   - Systems involved have unpredictably uncertain consequences.
   - The strategy was haphazard and ad hoc.
   - There are no specified Key Process Areas.
   - The lowest standard and the greatest danger.
2. Controlled maturity stage
   - Specifications are regulated.
   - Systems are organized and managed.
   - Programs are managed and executed in accordance with the arrangements that have been registered.
   - While this risk is smaller than the initial standard, it still persists.
   - The quality has improved since the beginning.
3. Specified maturity stage
   - Processes are well-defined and represented using principles, protocols, techniques, and software, among other things.
   - There is a medium level of consistency and risk involved.
   - System standardization is the main focus.
4. Quantitatively controlled maturity stage 4
   - Quantitative process efficiency and quality goals are identified.
   - Quantitative goals are dependent on consumer expectations, organizational needs, and so on.
   - Detailed analysis of process efficiency metrics.
   - Processes are of a better consistency.
   - Reduced chance.
5. Optimizing maturity stage
   - Processes and their efficiency are constantly being improved.
   - Progress must be gradual as well as creative.
   - Processes with the greatest efficiency.
   - Processes and their output with the lowest risk.

CMMI model—capability levels

A competency level is a set of unique and generic procedures within a particular process area that may aid the development of the organization's processes in that area. There are six capability levels for CMMI models with continuous representation, as defined below:

Capability level 0: incomplete
   - Unfinished procedure—completed partly or not at all.
   - One or two basic process area targets have not been met; no general goals have been defined for this standard, and this skill level is the same as maturity level 1.
Capability level 1: performed
   - It's possible that the process's success could fluctuate.
   - Performance, expense, and scheduling goals cannot be reached.

- A skill level 1 process is required to perform both unique and general activities for this level; however, this is just the first step in the process improvement process.

Capability level 2: managed

- The procedure is well-planned, tracked, and managed.
- Overseeing the process to ensure that the goals are met.
- There are model and nonmodel targets, such as expense, efficiency, and timeline.
- Use analytics to actively manage processing.

Capability level 3: defined

- A given mechanism is controlled and adheres to the organization's policies and procedures.
- System standardization is the main focus.

Capability level 4: quantitatively managed

- Mathematical and computational methods are used to monitor the procedure.
- Quantitative goals for process efficiency and output are established; process performance and quality are interpreted in mathematical terms and metrics.

Capability level 5: optimizing

- Insists on optimizing process efficiency on a continuous basis.
- Both subtle and innovative improvements are made to results.
- Places a premium on analyzing output data around the enterprise to identify and resolve common factors or problems.

4.8   GENERAL AGREEMENT ON TARIFFS AND TRADE

The GATT (General Agreement on Tariffs and Trade) is a free commerce agreement that has resulted in reduced tariffs and greater global trade. Between January 1, 1948, and January 1, 1995, the GATT controlled a significant part of international commerce as the world's first multilateral free trade pact. The agreement came to an end when the increasingly powerful WTO took its stance.

What was the GATT's function?

There were three major provisions of the GATT. The most important requirement was that each member provides the other most favored country status. When it comes to tariffs, both members must be treated similarly. There were no particular tariffs or customs unions between British Commonwealth countries. Where eliminating tariffs would cause significant damage to domestic manufacturers, tariffs were permitted. Second, the GATT made it illegal to limit the number of imports and exports. The following were the exceptions:

- Where a country has a surplus of agricultural goods.
- Emerging economies with nascent economies that needed to protect their balance of payments due to a lack of foreign exchange reserves.

Furthermore, for national security concerns, nations may implement trade restrictions. Among the topics discussed were patents, copyrights, and public morals. In

1965, the third clause, which deals with developing nations joining the GATT, was enacted. Developed nations agreed to eliminate tariffs on their imports to assist poor countries to flourish. Both rich and developing nations benefitted from lower tariffs. As the GATT increased the number of middle-class consumers across the globe, the market for developing country commerce grew as well.

- Member countries
    The initial 23 members of the GATT were Australia, Belgium, Brazil, Burma (now Myanmar), Canada, Ceylon (now Sri Lanka), Chile, China, Cuba, Czechoslovakia (now the Czech Republic and Slovakia), France, India, Lebanon, Luxembourg, the Netherlands, New Zealand, Norway, Pakistan, Southern Rhodesia (now Zimbabwe), Syria, South Africa, the United Kingdom, and the United States. The membership expanded to 128 countries by 1994. The meeting came dangerously close to spawning a third organization. It was going to be the International Trade Organization (ITO), a huge undertaking. The 50 nations that started the discussions intended it to be a UN body that would create rules regulating not just commerce but also housing, product agreements, business policy, foreign direct investment, and services. The ITO charter was agreed to in March 1948, but it was not ratified by the US Congress or any other countries' legislatures. The Truman Administration claimed defeat in 1950, effectively killing the ITO.

4.9   THE WTO–TRIPS AGREEMENT 1994

The fundamental principle of IP is that the result of human intelligence is secured in order for the owner of the intellectual production to profit from the outcome of their intellectual product or operation. To be protected, the intellectual object must be "owned" by anyone, that is, it must be "property."

The Accord on TRIPS of the WTO is the most comprehensive international IP agreement. It is critical in encouraging information and innovation commerce, resolving trade disputes over IP, and ensuring WTO members that they have the flexibility to pursue their domestic policy objectives. It situates the IP system in terms of innovation, technological change, and public benefit. The Agreement is a formal recognition of the significance of IP-trade relations and the necessity for a well-balanced IP regime. To comprehend the TRIPS Agreement, it is necessary to first study the IP system's background: What are the key types of IPRs? Why are these "powers" recognized? and How are they secured? Since the ratification of the first IP rules, these issues have been at the forefront of IP policy debates, and they continue to elicit heated debate. This module does not seek to outline various related legal and economic ideas, nor does it attempt to survey the variety of viewpoints presented in the debate; instead, it merely highlights some of the key principles and approaches. IPRs may be described as rights granted to individuals over their mental creations. They typically take the form of a restricted "exclusive right" given to a maker by national legislation to use the invention for a set period of time. A privilege like this allows the maker to prevent anyone from exploiting the production in such respects without his or her permission. The right holder will then derive economic benefit from the IPRs by using them personally or allowing someone to do so on his or her behalf. The digital world has also lent itself to trade in useful content covered by IPRs, and agreements on IPR licenses will form the foundation of commercial and technical collaborations. IPRs are tribal rights, which mean they are only applicable in the territories in which they were registered or obtained in some other way. To put it another way, the existence of a right in one country does not always imply the existence or legitimacy of an equal right in another (exceptions include systems of regional rights).

4.10   THE WIPO COPYRIGHT

The WIPO Copyright Treaty (WCT) is a unique agreement established under the Berne Convention to safeguard works on the Internet as well as the rights of their creators. The substantive provisions of the Berne Convention for the Protection of Literary and Artistic Works' 1971 (Paris) Act must be followed by any Contracting Party (even if it is not bound by the Berne Convention) (1886). Furthermore, the WCT mentions two copyright-protected subject matters: Computer systems, independent of their mode or type of expression; and details or other knowledge compilations (databases), in any form, that constitute intellectual inventions due to the collection or arrangement of their contents. (A directory that does not represent such a development is not covered by this Treaty.)

Apart from the privileges acknowledged by the Berne Convention, the Treaty further gives writers the following rights:

1. The right of distribution;
2. The right of rental; and
3. A greater right of public contact.

The right to circulate is the ability to make the original and versions of a work available to the public by selling or transferring ownership. The right of rental is the permission to allow the commercial rental to the public of the original and copies of three types of works: (i) computer programs (except when the computer software is not the primary focus of the rental); (ii) cinematographic works (but only in cases where commercial rental has resulted in widespread copying of such works, materially impairing the exclusive right); and (iii) literary works (but only in cases where commercial rental has resulted in widespread copying of such works, materially. The freedom to reach the public includes "the rendering available to the public of works in such a way that representatives of the public may access the work from a location and at a time personally chosen by them," as well as "the making available to the public of works in such a way that members of the public may access the work from a location and at a time personally chosen by them." On-demand, transparent contact over the Internet is the term in dispute. Article 10 of the WCT implements the so-called "three step" test for determining limits and exceptions, as defined in Article 9(2) of the Berne Convention, and extends its scope to all privileges. Such restrictions and exceptions, as defined in national law in accordance with the Berne Convention, can be applied to the digital world, according to the Agreed Statement that accompanies the WCT. Contracting States can create new exceptions and limitations that are more relevant to the digital world. If the requirements of the "three-step" test are met, existing restrictions and exceptions can be extended or new ones can be created.

In terms of length, any kind of work must be protected for at least 50 years. The exercise of the Treaty's privileges cannot be constrained by any formalities. The Treaty requires Contracting Parties to provide legal remedies for the circumvention of technological measures (e.g., encryption) used by authors in the exercise of their rights, as well as the removal or alteration of information required for the management (e.g., licensing, collecting, and distributing royalties) of works and their authors (rights management information). The Treaty requires each Contracting Party to take the steps required to ensure the Treaty's application, in accordance with its legal framework. Each Contracting Party must, in particular, ensure that compliance mechanisms are available under its law to allow effective action against any act of violation of Treaty rights. Such action would provide both immediate solutions to discourage infringement and long-term remedies to preclude further infringement. The Treaty creates an Assembly of Contracting Parties, whose primary purpose is to discuss issues related to the Treaty's upkeep and progress. It delegated the Treaty's regulatory functions to the WIPO Secretariat.

The Treaty was signed in 1996 and became effective in 2002. The Treaty is applicable to all WIPO members as well as the European Union. Such intergovernmental bodies may be admitted to the Treaty by the Assembly established by the Treaty. Instruments of ratification or accession must be deposited with the Director-General of the WIPO.

4.11   THE GLOBAL DIGITAL MILLENNIUM COPYRIGHT ACT 1998

The Digital Millennium Copyright Act (DMCA) of 1998 was the cornerstone of Congress's efforts to enforce US treaty commitments and modernize the country's copyright laws. However, as Congress understood, the only constant thing in life is transition. The DMCA's passage was just the beginning of Congress's continuing examination of the relationship between technical progress and copyright law in the United States. The DMCA needed the Report of the Register of Copyrights to support Congress with this ongoing phase.

In October 1998, the US Congress enacted the DMCA, which made major changes to the US Copyright Act. These changes were necessary in part to bring US copyright law into compliance with the WIPO's Copyright Treaty and WIPO Performances Phonograms Treaty, which are now managed by the WIPO. In the face of emerging digital information technology technologies, such as the internet, the DMCA improved the legal defense of IPR. The DMCA is divided into five parts, or divisions, the most immediate of which is Title II, which affects the Indiana University population. Title II lays out such legal obligations those Online Service Providers (OSPs) must follow in order to restrict their legal liability if a user of their service violates copyright laws. "An organization that offers the transmission, routing, or provision of connections for digital online communications" is what an OSP is known as. IU is considered an OSP for members of the IU information management system for the purposes of the DMCA.

According to the DMCA, an OSP must file with the US Copyright Office as an official agent. In the case of a possible copyright violation by someone whose OSP is IU, this agent is the appointed official to be contacted by a copyright holder. IU's registered DMCA agent is Sara Chambers of the Indiana University Information Policy Office (UIPO). The OSP is required to make information about the OSP's rules and general protocols for dealing with copyright infringement notices reasonably accessible to all users. When users' accounts and programs are discontinued due to persistent violations of copyright or other IP rules, the OSP is required to notify them. The copyright owner has the option of filing a lawsuit with the OSP for prosecution or personally serving legal notice on the infringer. The recipient would not be exempt from civil consequences if he or she claims ignorance of copyright infringement rules. It is the duty of the consumer to be mindful of these legal implications. As a result, IU actively advises users to inform themselves about the current status of copyright law as it pertains to file sharing over the Internet, as well as to stay current on amendments to copyright laws.

Title I, the "WIPO Copyright and Performances and Phonograms Treaties Implementation Act of 1998" implements the WIPO treaties. Title II, the "Internet Copyright Infringement Liability Limitation Statute" restricts the liability of online content providers that participate in copyright infringement. Title III of the "Computer Servicing Competition Assurance Act" creates an exemption for making a duplicate of computer software by allowing a computer to be used for maintenance or repair. Title IV contains five miscellaneous clauses dealing with the Copyright Office's responsibilities, distance education, the Copyright Act's exceptions for archives and ephemeral recordings, "webcasting" of sound recordings over the Internet, and the applicability of collective bargaining agreement commitments in motion picture transactions. The "Vessel Hull Construction Rights Act," or Title V, gives vessel hull designs a new degree of protection.

4.12   THE PRO-IP ACT OF 2008

On October 13, 2008, President Bush signed into law the Prioritizing Resources and Organization for Intellectual Property Act (Pub. L. 110-403). (Protection of Intellectual Property Act) The present measure strengthens the United States' IP laws in numerous ways. It amends Section 410 of the Copyright Act, for example, to formalize the registration infringement theory against the Copyright Office. The law also makes it unlawful (civilly or criminally) to distribute unauthorized copies of copyrighted works from the United States, including copies of phone records, and emphasizes that registration is not required for a criminal copyright prosecution. In addition, the bill modifies Section 506 of the Copyright Act to allow for the seizure of any property used to commit or promote a copyrighted works-related criminal offence. The Act creates a new regulatory framework for the government's anti-counterfeiting and anti-piracy efforts. For example, it creates an IPEC (Intellectual Property Enforcement Coordinator) inside the President's Executive Office. The coordinator will serve as the President's main spokesperson on local and international IP compliance issues, and this office will be responsible for developing and implementing a Joint Strategic Plan to fight counterfeiting and piracy. In addition, the Act provides additional funding for resources to investigate and prosecute IP breaches and other unlawful computer behavior, as well as enhanced investigative and forensic capabilities for the enforcement of IP laws.

In June 2007, the CACP released The Campaign to Protect America, a detailed set of recommendations to help our government in combating counterfeiting and piracy. Much of the Campaign to Protect America's proposals were included in the Prioritizing Resources And Organization For Intellectual Property Act Of 2008, or PRO-IP Act (P.L. 110–403), which was signed into law in October 2008 and marks a critical step in helping the federal government combat counterfeiting and piracy.

The CACP has managed to press the US government to take action. The following steps will be taken by the government to completely enact the PRO-IP Act:

- Implementation of the National IP Enforcement Strategy—In June 2010, the United States implemented the PRO-IP Act, which mandated the development of a national IP enforcement strategy. America's first-ever National IP Enforcement Strategy was announced by the IPEC, marking a significant move forward in our attempts to fight counterfeiting and piracy. Vice President Biden announced the initiative, and the occasion spoke volumes of the plan's relevance to the American economy. The IPEC released its Annual Report on Intellectual Property Compliance on February 7, 2011.
- Continued Federal, State, and Local Funding for Intellectual Property Enforcement—The CACP has helped in the funding of new FBI agents to investigate IP crimes after the PRO-IP Act was signed into legislation, as well as funding for US Attorneys to investigate them, as well as federal funds to support state and municipal law enforcement in combating IP violations. In September 2010, the Department of Justice reported about $4 million in grants to federal, local, and tribal IP enforcement agencies and prosecutors. We will keep pushing for funds for these projects.

The CACP will continue to collaborate with the Administration and Congress to ensure that the PRO-IP Act's provisions are fully enforced and financed.

4.13   ESSENTIAL COMPONENTS OF SOFTWARE DEVELOPMENT METHODOLOGY

A software development approach is a tried-and-true work procedure that helps system analysts, designers, project managers, and others to produce high-quality software in a managed and orderly manner. The tasks in the software development phase, as well as person and community obligations for completing them, are described by a technique. It also suggests specific techniques for carrying out the various functions, such as recording the logic of a computer application using a flowchart. A framework also provides guidance for monitoring software quality at different levels of growth. If a company develops such a technique, it is usually applicable to all software development projects the company works on (Figure 4.6).

Like for most situations, it's generally quicker and less expensive to prevent technical bugs in the first place rather than trying to repair the damage later. According to

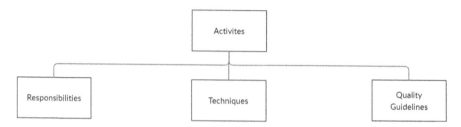

FIGURE 4.6 Components of software development methodology.

studies, identifying and eliminating a defect early in the product development process (requirements definition) will cost up to 100 times less than removing a defect in a piece of software that has already been shipped to consumers. If a flaw is discovered later in the production process, any rework on the deliverables created earlier in the process would be required. Thus, the costs of communicating and correcting the mistake would be higher. As a result, most software engineers strive to find and eliminate mistakes early in the development process, not just to save money but also to increase software quality. A product warranty lawsuit may be filed if it has underlying flaws that cause damage to the consumer. In two ways, using an appropriate approach will shield tech companies from legal responsibility. To begin with, an efficient approach decreases the amount of potential program errors. Second, incompetence on the part of a company is more difficult to establish if it practices universally agreed construction processes.

A good protection against a product liability lawsuit, on the other hand, will cost hundreds of thousands of dollars in legal fees. As a result, failing to carefully and reliably design applications will result in significant liability exposure. Quality assurance (QA) refers to approaches used during the production period to ensure a product's consistent service. In an ideal world, these approaches will be used at any step of the production process. Some software production companies, on the other hand, consider testing to be their only QA tool since they lack a standardized, standard approach to QA. Rather than searching for defects in the production process, such businesses focus on monitoring only before the product ships to ensure accuracy.

4.14   SOFTWARE TESTING

Software testing is the method of evaluating the functionality of a software program with the goal of determining whether the designed software meets the required specifications and identifying bugs in order to deliver a better product. Subroutines or programs are the building blocks of software. Large structures are formed by combining these units. In a procedure known as dynamic testing, the code for a finished unit of software is tested by directly entering test data and matching the results with the predicted results.

- Manual testing is the process of manually evaluating software in order to learn more about it and determine what is working and what is not.
- This usually involves verifying all of the functionality listed in the specification documents, but it also often entails testers putting the software through its paces from the perspective of its end customers.

- Manual test plans may vary from completely scripted test cases with precise instructions and expected results to high-level guides that guide exploratory study sessions.
- Automation testing: Automation testing is the method of finding bugs in software by using an automated tool.
- Using automation software, testers run test scripts and produce test results automatically in this process.

Dynamic testing can be divided into two types: black-box testing and white-box testing.

Black-box testing entails seeing the software unit as a system with predicted input and output actions but unexplained internal workings (a black box). The unit passes the test if it exhibits the predicted behaviors with any of the input data in the test suite. Black-box checking is done without the tester knowing something about the code's composition or design. As a result, it is often performed by someone other than the code author.

- Unlike black-box testing, white-box testing considers the software machine as a system with planned input and output behaviors but unknown internal workings. White-box research entails thoroughly understanding the logic of the program unit and testing all feasible logic pathways through it. Each program statement must be executed at least once, so the test data must be carefully assembled.
- Gray-box testing is a hybrid between White-Box and Black-Box testing. This method of research necessitates access to concept documents by the tester. This aids in the development of better test cases during the process.

4.15 PROCESS AND PRODUCT QUALITY

Product:

The end result of the software development process is the product. A commodity is created based on the needs and demands of the user.

Process:

A procedure is a series of steps that must be taken in order to produce a product. A procedure is a blueprint that can be used to make a variety of things in the same way (Figure 4.7).

The following are a few main distinctions between Product and Process.

4.15.1 QUALITY ASSURANCE AND STANDARDS

Quality is difficult to define, but it can be summarized as "fit for use or purpose." It's all about satisfying consumers' desires and desires in terms of product functionality, style, efficiency, longevity, and price. Assurance is nothing more than an optimistic statement about a good or service that inspires confidence. It is the assurance that a good or service can perform well. It ensures that the product will perform flawlessly and in accordance with the standards or specifications. Software testing QA

FIGURE 4.7 Quality assurance process.

S. No.	Key	Product	Process
1	Concept	The end result of a production period is the commodity.	A procedure is a series of procedures that must be taken in order to produce a product.
2	Focus	The emphasis in product creation is on the end result.	The procedure focuses on each move that must be taken throughout the implementation of a software product.
3	Life	The life span of a commodity is usually small.	The life cycle of a method is usually very long.
4	Goal	The primary aim of product production is to accomplish the mission and execute the product effectively.	The primary aim of a method is to manufacture high-quality materials.

is defined as a method for assuring the quality of software goods or services that a business provides to its customers. QA is concerned with making the software development process more dependable and competitive while adhering to the quality standards set for software products. QA testing is a common term for QA.

In QA, the PDCA cycle, commonly known as the Deming cycle, is a fixed cycle. Plan, do, check, and act are the four stages of this cycle.

These processes are repeated on a regular basis to guarantee that the organization's operations are evaluated and improved. Take a look at the QA in the image above. An in-depth look at each step of the procedure is as described below:

- Plan and create process-related objectives, as well as the processes required to deliver a high-quality final result.
- Perform: Develop and test processes, as well as "do" system changes.
- Check: Procedures are monitored, changed, and checked to confirm whether they meet their objectives.
- Act: A QA tester may take the necessary measures to improve processes.

An organization uses QA to verify that the product is developed and delivered properly. This helps to reduce the number of problems and faults in the final product.

4.15.2   QUALITY PLANNING

Quality planning is the process of establishing a project's quality schedule. The product quality plan outlines software quality criteria and how they will be assessed. The quality strategy establishes which operational criteria apply to a certain product and manufacturing process. A quality plan is made up of the following elements:

1. Product introduction.
2. Product strategies.
3. Detailed explanations of the processes.
4. Quality objectives.
5. Risks and risk mitigation.

The quality plan includes a description of the app's quality assessment methodology as well as the app's most important quality characteristics.

4.15.3   QUALITY CONTROL

Throughout the software development process, quality management ensures that QA procedures and practices are followed. The outputs from the software development process are compared against the project requirements in the quality management process. To guarantee the quality of software project outputs, daily product evaluations and/or automated software evaluation may be utilized. A network of individuals conduct QA assessments. They examine the software and software procedure to ensure that the project guidelines are met and that the documentation and records are compliant. Automated software evaluation examines the software using a program that compares it to the implementation project's requirements.

4.16   SOFTWARE QUALITY AND ASSURANCE TECHNIQUES

The appropriateness of a software product for its intended purpose determines its consistency. That is, a high-quality product fulfills the expectations of its consumers. The suitability of software items for usage is typically described in terms of fulfilling the SRS guide's requirements. "Fitness of purpose" is an acceptable understanding of quality for specific equipment, such as a car, a table fan, a grinding machine, and so on. However, "fitness for purpose" is not an entirely acceptable definition of software quality.

Many quality methods are included in today's knowledge of software product quality, including the following:

If a digital interface can operate on a range of operating systems, on various computers, with other software devices, and so on, it is called scalable.

Usability: A software interface is more usable if all types of consumers can conveniently invoke the system's features.

Reusability: If separate components of a software system can be easily replicated to produce new applications, it has outstanding reusability.

Correctness: A software product is right if the SRS document's different specifications have been applied properly.

Maintainability: A software system is maintainable if glitches can be easily fixed when they arise, new tasks can be easily applied to the product, and the product's functionalities can be easily changed, among other things.

System for managing software quality

A quality control scheme is one of the most common ways for businesses to ensure that the goods they produce are of the highest possible quality.

The below are the components of a quality system:

Individual responsibilities and managerial structure: The entire organization is responsible for a quality scheme. Any organization, on the other hand, has several quality departments that conduct different quality system activities. The arrangement's quality scheme should have the approval of the top management. Some employees would take the quality system negatively if the quality system is not supported at a high level in the business.

Activities in the quality management system: The functions of the quality system include the following:

- Project auditing
- Quality system review
- Development of principles, procedures, and protocols, among other things
- Production of publications for senior management summarizing the success of the organization's quality system.

REVIEW QUESTIONS

1. What is open-source code? Explain fundamental premise behind it.
2. What are the four most popular forms of product liability lawsuits concerning software?
3. Explain the key issues involved in intellectual property rights.
4. Write a note on CMMI.
5. Explain development of safety critical system.
6. What is intellectual property right? List all types of intellectual property right.
7. "The open-source coding scheme is detrimental to software developers." What is the truth or falsity of the statement? Justify your answer.
8. Explain various software quality assurance techniques.
9. Differentiate between process and product quality in software.
10. What do you understand by WTO Trips Agreement 1994?

MULTIPLE CHOICE QUESTIONS (MCQs)

1. **The usage of knowledge and concepts that are of value is protected under intellectual property rights (IPR).**
   a. Commercial value
   b. Social value

c. Moral value
d. Ethical value

2. The term "Intellectual Property Rights" covers

a. Know-how
b. Trade dress
c. Copyrights
d. All of the above

3. Which of the following privileges cannot be transferred or licensed to another?

a. Patents
b. Trademark
c. Designs
d. All of the above

4. Which of the following items may be protected by a patent?

a. Composition of matter
b. Machine
c. Process
d. All of the above

5. In "quid-pro-quo," quo stands for

a. knowledge disclosed to the public
b. monopoly granted for the term of the patent
c. unique right to develop, market, and use the innovation
d. none of the above

6. Geographical Indication belongs to

a. intellectual property right
b. community right
c. private right
d. both (a) and (b)

7. Which of the below is/are used in Goods Geographical Indications?

a. Manufactured
b. Foodstuff
c. Handicraft
d. All of the above

8. The phrase "legally enforceable arrangement" applies to an agreement that is enforceable by statute is called as_____

a. Unenforceable agreement
b. Void agreement
c. Illegal agreement
d. Valid agreement

9. In India, literary works are covered before they are released at

a. Sixty years after the death of author
b. Thirty-five years after the death of author
c. Twenty years after the death of author
d. Lifetime of author

10. **Which of the following is not a technique for software quality assurance?**
    a. Usability
    b. Reusability
    c. Maintainability
    d. Redundancy

ANSWERS TO MCQs

Q1: a, Q2: d, Q3: b, Q4: d, Q5: b, Q6: d, Q7: d, Q8: d, Q9: a, Q10: d

5 Ethics of IT Organization

LEARNING OBJECTIVES

- Grasp the field of organizational behavior and its use in the workplace.
- Describe the strategies for fostering ethical conduct as well as the regulations that promote ethical business operations.
- Discuss acceptable communication strategies and styles in the workplace.
- Examine how organizational transformation affects individuals and the workplace.

5.1 INTRODUCTION TO ETHICS OF IT ORGANIZATION

Organization ethics refers to how an organization can react to the external world. Organizational ethics refers to a set of rules and standards that determine how employees can act in the workplace.

It also applies to the code of ethics adopted by workers of a specific company. Any business exists to make money, but the manner in which it does so is more important. No company should focus on shady tactics to make profits. It is vital to remember that money is not the most valuable factor: pride and honor are much more important. A person's first goal might be to make money, but he or she should not stoop to that extent just to do so. Children under the age of 14 years must not be working in any company. Childhood is the most precious time in one's life, and no child should be denied it. Employees do not kill or exploit data in order to accomplish their objectives. In the business world, tampering with data is considered dishonest and unprofessional. Bear in mind that if one is sincere, things will still go in his or her favor.

Employees should not divulge any of the company's confidential information to third parties. Do not divulge any of the company's rules or recommendations to anyone. It is preferable to avoid debating official affairs with friends and family. In no condition can classified details or information be leaked. In monetary transactions and all forms of trade, total justice is required. Organizations are forbidden from discrimination against workers depending on their identity, physical appearance, age, or family history. Employees who are female must be handled with dignity. Do not expect your female workers to work late. Discriminating against workers because they do not come from a rich family is immoral. Employees should be judged purely on the basis of their work. Employees must not be used by the company. Employees must be compensated for their contributions and hard work. When people are working late at night, make sure they are credited for their time. Employees must collect their arrears, salaries, incentives, and other reimbursements on schedule, according to administrators (Figure 5.1).

DOI: 10.1201/9781003280989-5

FIGURE 5.1 Organizational ethics.

It is illegal to rob office premises. Employee protection must be prioritized by the company. Individuals should not be placed in dangerous situations. Making false claims to customers is unprofessional. The substance must be distinctly represented in the advertising. Don't make any commitments that the business can't keep. To expect loyalty from your clients, you must be honest with them. It is completely dishonest to deceive your clients.

5.1.1 RULES AND LAWS OF IT ORGANIZATIONS IN THE UNITED STATES

The electronic protection laws and privacy structure in the United States are probably the world's largest, most comprehensive, and effective. The privacy mechanism in the state is more dependent on ad hoc government compliance and private prosecutions. Cyber security policy currently consists of executive branch guidelines and legislative law that covers information technology and data networks.

- Control by the federal government
  The following are the three primary federal cyber security laws:
  - Health Insurance Portability and Accountability Act (HIPAA) of 1996
  - Gramm–Leach–Bliley Act of 1999
  - The Federal Information Security Management Act (FISMA) was included in the Homeland Security Act of 2002
    These three rules require healthcare organizations, financial institutions, and government departments to safeguard their processes and

data. These laws, however, are not foolproof in terms of data protection and only require a "fair" degree of security. FISMA, for example, "requires the establishment and application of mandatory laws, principles, practices, and recommendations on information security" and extends to all federal agencies. However, certain computer-related businesses, such as Internet Service Providers and tech firms, are not covered by these laws. Furthermore, the regulations' ambiguous wording leaves a lot of space for understanding.

- Federal rules in the current years

   The federal government is proposing several new cyber security regulations as well as amending existing ones in an attempt to enhance the security environment. Here are a few examples:

   - Act and Exchange Cyber Security Knowledge (CISA)

      Its goal is to strengthen cyber security in the United States through enhancing knowledge exchange regarding cyber security risks, among other things. The law mandates that the US government, as well as technology and manufacturing companies, share information about Internet traffic. On July 10, 2014, the bill was presented in the US Senate, and on October 27, 2015, it was approved.

   - The 2014 Cyber Security Improvement Act:

      It became law on December 18, 2014. It is a voluntary public–private collaboration that seeks to enhance cyber security by bolstering cyber security research and growth, talent development and education, and public knowledge and preparedness.

   - The 2015 Federal Exchange Data Breach Disclosure Act:

      This bill mandates that a health insurance exchange alert any customer whose personal information is known to have been obtained or accessed as a result in a security violation in any system operated by the exchange as soon as possible, but no later than 60 days after the breach is discovered.

      The National Cyber security Safety Development Act of 2015 is a federal statute that was passed in 2015. This bill amends the Homeland Security Act of 2002 to allow the National Cyber Security and Communications Integration Center of the Department of Homeland Security to incorporate tribal governments, research centers and intelligence exchange, and private organizations as nonfederal members.

5.1.2   Rules and Laws of IT Organizations in Other Countries

Countries such as Australia and New Zealand enacted national freedom of information (FOI) laws as early as 1982 (prior to the recent round of FOI laws being adopted around the world), and then went on to amend their legislation to get their FOI regimes up to date. Furthermore, if one looks at South Asia, India's national FOI law (enacted in 2005) is often quoted as an example of the role that grassroots organizations can play in the implementation of such legislation, as well as the effects that exercising FOI can have on people's lives. Nonetheless, India's history shows the challenges that can emerge during the implementation process, including institutional opposition, public officials' capability deficiencies, a lack of public knowledge, and the need to strengthen compliance with constructive transparency obligations and ensure conformity with other regulations (for assessments on the case of FOI in India, you can read here, here, and here). In turn, FOI laws have recently been enacted or entered into effect in countries such as Mongolia (a FOI law was passed in 2011), Bangladesh (legislation has been in force since 2009), and Indonesia (a FOI law has been in force since 2010), and draught FOI laws are currently being debated in a host of others.

China has not adopted and will not enact a single regulation that includes all facets of data security. The Standing Committee of the National People's Congress reported in December 2019 that a Personal Information Privacy Law and a Data Security Law will be drafted in 2020. This is largely attributed to China's need for its data security system to safeguard both personal data and online privacy, which are important to private users, as well as "important data" that are of governmental importance. In an age of globalization, this dual-purpose policy was conservative. Given the recent breakdown of globalization, as illustrated by increasing populism in many big Western countries, it does not seem to be so conservative after all.

Regulations and specifications on data security are distributed through different regulations, as well as certain prescribed operating criteria at the national level, under the existing framework. China's data security legal frameworks are divided through the four fields mentioned below:

1. Specifications for general data protection, such as the Cyber security Legislation and its implementing laws and standards;
2. Data security for national secrets, including the Legislation on the Protection of National Secrets and its implementing laws and rules;
3. Personal information security, including the Civil Code, the Consumer Rights and Preferences Act, and applicable national standards; and
4. Essential data security laws, such as those provided by ministries or municipal councils that refer to a particular sector or country.

5.2   ELEMENT OF AN ETHICAL ORGANIZATION

Recently, there has been a lot of debate on corporate ethics. Of late, the bulk of the focus has been on the lack of corporate ethics. Because the uncertainty of day-to-day life obscures your company's philosophical side, ethical conduct is often ignored by entrepreneurs. But have no fear: Reconnecting to a sound ethical foundation is much better than doing an "ethics transplant."

Ethics is viewed by many business practitioners as if it were a sermon on the mount. While it's impossible not to lecture when it comes to ethics and morality, I've attempted to define several points to remember, as well as guidelines for ethical conduct. It's exciting to understand how good corporate behavior represents fair company practices, but maybe that's why great businesses are great in the first place.

5.2.1 RESPECT

As a company owner, you must respect yourself and associate yourself with others who will respect you. Note that getting a high degree of appreciation does not indicate that you should work on autopilot. While you can assume that your employees will do their jobs to the best of their abilities, they will require coaching, training, and direction. However, respect and trust will make it easier for you to avoid micromanaging them. People who don't respect, or who don't respect you, should not be employed or do business with you. These are the people who don't value their coworkers, clients, vendors, or themselves in the end. Take steps as current partnerships deteriorate. Make every attempt to restore shared respect; but, if it is no longer practicable, let the individual go.

5.2.2 HONOR

Good ethics are founded on the basis of good citizens. They're still excellent champions for doing the right thing. Pay particular attention to top performers and those who exemplify the company's philosophy. The bulk of firms honor high achievers and performers. Don't limit yourself to caps and revenue numbers. Individuals who demonstrate exceptional conduct and have made contributions on their behalf should be remembered and praised. There are individuals who have supported you in being successful, and you must publicly and privately appreciate and respect their efforts.

5.2.3 INTEGRITY

It's hard to resist appearing preachy or parental when it comes to integrity. Do not trick, rob, or defraud. Make your word your bond and respect it at all times. When you make a mistake, confess it and make good on your word. Treat people the same way you'd like to be handled. People who lack credibility should not be hired or retained. They would be distrusted by their peers, clients, and vendors. The lack of confidence spreads like a plague, and they will gradually lose faith in you. Ascertain that no one is manipulating the company's principles to make a fast buck. Making a poor bargain to fulfill a quota or target is, after all, not just dishonest, but it's very sometimes unprofitable in the long run.

5.2.4 CONSUMER FOCUS

Without consumers, a business is nothing. More to the point, a corporation has no value if it does not deliver what customers want and are willing to pay for. The responsibility you have to the market is reinforced by a focus on your customers. Your decisions have an effect on your staff, clients, associates, and, eventually, consumers. It is part of the ethical duty to serve all of these people. Selling your clients short not only jeopardizes your integrity, but it also jeopardizes your company's long-term viability.

5.2.5 RESULT-ORIENTED

You wouldn't be an entrepreneur if you weren't already concentrating on outcomes, but ethics play a role in outcomes as well. Don't try to get results at all costs. Act to meet your goals while remaining true to your company's standards. Results should be obtained by making something that consumer's desire and manufacturing and distributing it at an expense that is equitable to all parties concerned. Good managers specifically mention the outcomes they want and help and assist their workers in achieving those outcomes. They provide performance reviews in order to help the employees meet their full potential and deliver the outcomes that the organization wants to survive. Results are more than just figures in a successful (and ethical) company. They serve as future benchmarks and lessons learned, as well as current priorities.

5.2.6   RISK TAKING

So far, you may have had the idea that responsible firms are tentative and mousy, afraid to do the wrong thing. That obviously isn't the case. Risk-taking is how organizations compete, succeed, and expand. They don't take the safe path. Great firms innovate, work beyond the box, and execute new ideas. They reinvent themselves and reward those who are able to take chances. Risk-taking isn't a challenge to your integrity as long as you hold to your philosophical values. Employees who are able to take calculated risks are drawn to great businesses, which inspire, sustain, and reward them. They share the benefits with those who deliver when the risks pay off. When the threats do not pay off, they evaluate what went wrong and find out how to do it differently next time. Consider this: When you're taking chances, would you rather be surrounded by people you know and admire or sharks and snakes?

5.2.7   PASSION

Great companies are made up of individuals who are passionate about what they do. There are people who are working for you because they want to feel the excitement, not just to get paid. They are enthused, motivated, and confident that their actions and activities would have an impact. People who don't have a raging passion for what they do put in little work, get paid, and go home. When you can come in late and leave early, why work too hard? These people are role models for others. People can express their enthusiasm in a lot of ways, but be mindful that putting in extra time on a job or working on the weekend can show just as much excitement as energetic cheerleading.

5.2.8   PERSISTENCE

People in excellent companies have a deep ability to keep working. And if the results aren't what they planned or if consumers hesitate to shop, they'll keep going. Their perseverance stems from a love for what they do and a conviction that this community of individuals, this business, has the best chance of "making it" of any they could join. As a result, they bring in more time and want to take chances. They are honorable and trustworthy. They stay focused on the customer's desires and needs. And they aren't happy until they meet the desired goals and outcomes. As the chief, you would dedicate a considerable amount of time and attention to identifying others who share these beliefs. Discuss the role of these principles in your team's policy, schedules, and decisions. You must draw a straight distinction in the business between "what is allowed" and "what is not allowed." When anyone crosses the line, the leader must warn them of their transgression. Give them another chance to change their behavior, or let them go, depending on the person (and the incident). It is unethical to take no action. Following your values can be the height of ethical conduct. Not unexpectedly, it not only makes legal sense, but it also makes sound economic sense.

5.3   NEED OF TRADITIONAL AND NONTRADITIONAL WORKERS IN ORGANIZATIONS

The number of undergraduate degrees granted in computer science, computer engineering, and information technology at doctoral-granting computer science schools in the United States and Canada has fallen significantly, according to the Computing Research Association. The number of undergraduate degrees, on the other hand, has increased to about 15,000 in recent years. The federal government's prediction of increased demand for jobs in computer science-related fields led to the rebound. According to the Bureau of Labor Statistics, there were 3.4 million workers working in IT-related occupations in the United States in 2010, and the industry is expected to create almost 750,000 additional jobs between 2010 and 2020. Because of the reduction in undergraduate degrees awarded in computer science and engineering sectors, IT companies and organizations that use IT goods and services are concerned about a lack of jobs in the United States to fill not just the anticipated new vacancies, but also the many people who would withdraw from current positions.

Employers are gradually looking to nontraditional outlets to find IT employees with the expertise they need, such as temporary workers, H-1B workers, and outsourced overseas workers, when they face a possibly long-term lack of skilled and experienced workers. Employers must make ethical choices on whether to hire new and more professional employees from these outlets or grow their own staff to fulfill their business's demands as they weigh these options. Many firms rely on temporary staff, independent contractors, and consultants in addition to conventional jobs.

* Individually or through a temporary staffing or consultancy service, these employees can interact with your organization.
* You will stay for as little as one day or as many years.
* Provide a range of services, from clerical support to technically professional project work.

For so many factors, it's no surprise that many human resource (HR) specialists have difficulty finding out who, when, and how much of their nonemployee employees they have. Although temporary labor has been used by employers for decades, an analyst believes that today's "nonemployee" population is greater and more complex. HR practitioners, too, have trouble keeping up with the terminology: Although many people do call this category of staff "contingent," others argue that the word doesn't adequately reflect contractor and company partner partnerships.

The ways in which each form of labor is paired with physical resources as part of the production process may play a role in the productivity of nontraditional workers in contrast to traditional workers. Technological innovation can alter the benefits businesses derive from nontraditional work by altering the characteristics of resources. Some also hypothesized that the advent of new technologies would raise the need for contract jobs, such as labor hire employees. Nontraditional occupations are positions that have historically been held by only one demographic. Nontraditional Jobs, according to the US Department of Labor (DOL), are occupations in which people of one gender account for fewer than 25% of the total workforce.

It's a smart idea to think about all of the choices before settling on a career path. But don't restrict your career opportunities because you feel the job you want is just for a certain gender. Men and women frequently have preconceived views of what constitutes women's work and what constitutes men's work. There are few occupations that have conditions that essentially restrict opportunities to men and women. Nontraditional jobs are an opportunity that many people ignore while looking for work. Tradition and the way we've been socialized have long shaped our job choices; as a result, men and women are frequently ignorant of the diversity of opportunities open to them. The aim of identifying nontraditional options is to increase participation in and access to these careers for women who are underrepresented in these fields.

5.4   GAINING VISIBILITY AND STANDARDIZATION

The recruiting business has been turned on its head by developments in the employer–employee relationship. Talented employees are reconsidering the role of employment in their lives, and many are finding a work–life balance. Many applicants have lost confidence in working with a specific company on a long-term basis. Instead, they have adopted the "gig economy," which helps them to work on their own schedules. They decide what they want to work, what projects they want to take on, and which businesses will help suit their immediate needs.

Managing temporary staff is difficult for mid to big firms. Frequently, each agency takes its own decisions, resulting in differences in pay for the same jobs. Some contractors may often find new administrators who lack the skills and abilities required to meet the specific demands of a nontraditional workforce. With the rise of social media, a single negative experience by a contractor will quickly tarnish a company's talent image. Increased exposure is the alternative to contingent employee risks. Consider a company-wide approach instead of delegating contractor recruitment and administration to each department. All matters pertaining to contingent employees should be centralized, including the recruiting process, deciding wage grades, and integrating them into the workflow. You should expect more insight into the role contingent employees perform in the market as a result of centralization, ensuring that executives and HR experts have oversight over these critical processes.

Standardization's advantages:

Standardization essentially ensures that the workers can follow a well-established, time-tested protocol. Standardization, when performed right, will eliminate uncertainty and guesswork, ensure consistency, enhance efficiency, and boost employee morale.

The following are some of the advantages of standardization:

- Increases transparency—A standardized method eliminates the need for guesswork or further digging.
- Ensures consistency—When work is completed in a predetermined, efficient manner.
- Improves productivity because the workers won't have to ask about or sift through papers to find answers.
- Increases staff productivity by encouraging workers to take pride in mastering the process and honing their skills.
- Improves customer support by ensuring that each ticket is managed as quickly as possible.

By eliminating inefficiency, standardization increases quality. This is the product of reducing complexity and maintaining quality management: Tasks are performed more quickly, and there are less quality control problems as a result of tasks that were not completed properly the first time.

5.5   CONTINGENT WORKER

A temporary worker is someone who is employed on the basis of a short-term need rather than a fixed contract that determines long-term jobs with an organization. These workers work with the company for as long as they are required. Their employment may be temporary until a certain assignment is done, or they could be employees who have mentioned that they do not intend to be working for long. Contingent employees include those who work on construction projects, those who set up phone and Internet networks for new companies, and those who only work during certain seasons. Temporary employee, seasonal employee, and contract worker are all words that are related.

Contingent staff are persons who work on a mission or on a contract basis. This mark refers to all employees, regardless of their skill set or level of experience. Temporary employees given by an outside recruiting firm of freelancers and independent contractors make up this workforce. Contingent jobs, according to the Bureau of Labor Statistics, are a working situation in which a person does not have an explicit or implied arrangement for long-term employment. Independent contractors, migrant employees working through job brokers, on-call or day laborers, and on-site workers whose services are delivered by contract companies make up the transient workforce. When a company's technical manpower demands fluctuate greatly, dependent IT personnel are likely to be used. As contractors for an internal transformation initiative, strategic advisors on a product creation unit, and supplemental personnel on a variety of other short-term tasks, such as the implementation and construction of new information systems, workers are often employed on a temporary basis. These workers usually join a team of full-time staff and other contract workers for the duration of a project before going on to their next assignment. The company's desire for them determines whether they work, where they work, and how well they work. They have no contract for continued work, either overt or implied. Temporary recruiting agencies and employee-leasing businesses are good places to search for temporary employees. Job seekers in a wide variety of job categories and ability levels are hired, trained, and evaluated by temporary staffing agencies, who then appoint them to clients as appropriate. Temporary jobs are often used to offset employee absences and injuries, monitor seasonal workloads, and assist with special events.

They are not, however, considered official employees of the corporation and are thus ineligible for perks such as vacation, sick leave, and prescription care. Temporary jobs are also paying a higher minimum rate than full-time staff performing similar work and they do not earn extra pay from employer benefits. People who want full consistency in their job schedule as well as a variety of work opportunities can find temporary working arrangements appealing. Other staff consider irregular job assignments because they are unable to secure permanent jobs. In employee leasing, an organization (referred to as the subscribing firm) moves all or part of its employees to another company (referred to as the leasing firm), which oversees all HR-related operations and expenses, such as payroll, recruitment, and benefit administration. These staff are rented by the subscribing company, but they remain employees of the leasing firm. Employee-leasing companies work with a small administrative, distribution, and marketing workforce to keep overhead down and move the benefits on to their customers.

Employee leasing is a type of co-employment where two contractors share legal rights and responsibilities with the same employee or group of employees. Unique laws apply to employee lease agencies in respect of employees' compensation and unemployment benefits. Since the contractors are legally employees of the rental corporation, they could be liable for some of the company's benefits. Using a consultancy company to hire temporary IT staff is another choice. Professionals with a wide range of knowledge and experience, up to and including world-renowned industry experts, work for consulting firms; as a result, these firms will frequently have the precise qualifications and experiences that a company requires for a specific project. Consulting companies collaborate with their customers on engagements with well-defined planned objectives or deliverables (e.g., creating an IT strategic plan, implementing an enterprise resource planning system, or choosing a hardware vendor are all things that need to be done). The length of the agreement and the rate of pay for each of the contractors, who are led on the engagement by a senior manager or director from the consultancy company, are usually specified in the contract with a consulting firm.

5.5.1 RIGHTS OF CONTINGENT WORKERS IN THE IT INDUSTRY AT GLOBAL LEVEL

Over the last 50 years, the UK economy has moved significantly from comparatively steady employment in the public or private sector to a stronger dependence on contractual or contract labor. The on-demand or transient workforce (or gig economy) has expanded as the industrial sector has shrunk. Highly trained technicians and contractors can be employed in almost every sector as part of the contingent workforce. To fill their workforce, major companies are increasingly recruiting more versatile, temporary employees. In the United Kingdom, the contingent population is steadily expanding and shows no signs of slowing down.

Demand has powered the global transient workforce, with both companies and employees benefiting from the versatility it offers. However, companies, unions, employees, and regulators can face difficulties as a result of the use of contingent workers. If market models expand and the number of freelancers, contractors, suppliers, temporary agency, on-demand, and gig employees rises, pressures on the conventional employer–employee model will result in lawsuits, protest, and regulatory reform.

Since the dangers of misclassifying nonpermanent employees, especially the self-employed, are evolving, all employers employing temporary workers, not just gig employers, need to be mindful of these new trends. However, given the persuasive advantages of mutually beneficial versatility of both managers and employees, these trends are likely to intensify rather than fade with time.

Although policymakers around the world support the emergence of new ways of working that help fuel economic growth and are wary of limiting them excessively, some are responding by concentrating on two issues: How to defend lower paid contingent jobs by improving their rights and how to fix tax shortfalls associated with the increase in self-employment. In the United Kingdom, for example, the government has announced that it would keep employers more responsible for determining the submission of, and possibly paying, job taxes as they appoint allegedly "self-employed" contractors, and recent EU law provides basic protections for those who work erratic schedules.

Concerned over a drop in tax and social security payments, as well as ongoing misclassification lawsuits, governments and regulators have introduced new legislation to address the use of contingent jobs. Following a high-profile misclassification case law in 2018, California recently committed on landmark laws to decide whether a worker is an employee or an independent contractor. In 2020, other states in the United States are expected to focus on the definition of temporary employees. This activity is unlikely to be consistent, with some politicians predicted to support industry by classifying some temporary jobs as independent contractors rather than staff, whereas others are expected to take a more protectionist approach.

Canada, Australia, Ireland, the European Union, the United Kingdom, Finland, Russia, the Czech Republic, Germany, Italy, France, and Romania are all considering or have already adopted policy reforms. These vary depending on the region, but normally involve one or more of the following: greater limits on the use of contract agency, fixed-term, or zero-hour employees; expanded unemployment social contributions for fixed-term workers; additional scheduling rights, including minimum advance notice for shifts; minimum or fair wages for certain contract workers in contrast to their employee counterparts; increased state workplace checks or enforcement; prejudice and intimidation.

When using contract employees around the world, consider

- Performing a company requires audit;
- Following agency worker (leasing), fixed term, and all other special contingent worker laws;
- Receiving guidance when writing conditions of engagement, keeping them updated, and making sure they're met in operation;
- Keeping them separate (don't incorporate them into the organization).

Keep up with case law and regulatory amendments including misclassification.
Do not

- Re-engage temporary workers for long periods of time;
- Treat them as permanent staff;
- Exercise day-to-day authority over them, including instructing when, where, and how they work;
- Allow them to become economically dependent or incorporated into the workplace.

5.5.2 Pros and Cons of Contingent Workers

Using temporary staff provides a lot of possible benefits. The following are a few of the more common advantages:

1. Bridging the competency divide

   Many companies are facing a skills gap and are having trouble recruiting staff with unique skill sets. Contingent jobs have a large talent pool from which to select applicants with the unique expertise that permanent employees need for a project. Let's imagine the marketing department is searching for someone to work on a project that involves graphic design expertise. Rather than recruiting anyone, you might rent out the aspect of the project to a freelancer.

2. Adaptability

   Contingent employees will help you manage and change your staffing ratios. They allow companies to adjust rapidly to changing economic and consumer conditions without having to lay off permanent employees.

3. New points of view

   New experiences are taken about by seasonal staff. They have the opportunity to alter the way some facets of the company work, allowing for greater change and growth. Permanent workers can result in a workforce that lacks imagination and creative thinking.

4. Cost-cutting

   A contingent workforce enables businesses to pay for services only when they are required. It lowers the costs of hiring, on boarding, and preparation that come with permanent hires. Benefits, vacation, and sick leave are not included in the expenses of contingent jobs.

Contingent workers' cons (disadvantages)

Using temporary staff is rife with problems. The following are some basic problems and how to solve them:

1. Contact organizations with temporary staff face a big communication problem. Contingent staff and permanent employees should be handled similarly.

2. Lack of understanding of employment arrangement

A lack of awareness of how temporary workers' job conditions vary from those of permanent employees can create problems. A partnership between two parties is referred to as a contractor arrangement. Contingent employees are either self-employed or employed by a vendor. Discipline may be a challenge when disciplinary problems emerge and are often responsible to the organization. Managing temporary workers need not be that troublesome if success and behavior expectations are explicitly conveyed to the worker and/or vendor."

3. Management/visibility gap

Without a cohesive approach, department heads are often charged with overseeing their own temporary staff. This leads to disengaged unsupervised jobs, a shortage of exposure, and higher costs. To escape this stumbling block, you must

- build a comprehensive strategy that puts all teams together;
- not let teams build their own management practices;
- build an integrated contingent management approach that is concentrated in one area.

This eliminates inconsistencies in payment systems as well as time-consuming manual tasks. It also enables more worker awareness, faster billing, enforcement, and worker review. Another way to reduce fragmented management and improve visibility is to use a vendor management system (VMS), which is cloud-based and central for managing a company's transient workforce. Contracts, expenses, time logs, recruiting, on boarding, compliance, and other information pertaining to contingent workers are all visible through a VMS.

Recruiting and screening are the fourth and fifth steps in the recruiting process.

Another issue is that, in general, temporary hires have poorer screening standards than permanent employees. Contingent staffs often have access to the same business services and confidential information as permanent employees. Employee theft, fraud, computer security violations, lack of enforcement, litigation costs, and a weakened reputation will all stem from weaknesses in screening procedures. Recruiters and organizations must establish clear background check procedures that identify the forms of screening that will be performed for the positions they are applying for. Since organizations may be at risk of recruiting migrant workers who do not have legal authorization to work in Canada, multinational screening processes should be included.

5.5.3 Differences between Contingent Workers and Employees

It's not as straightforward as running an advertisement online or putting up a flier to fill a hole, whether it's for seasonal support or some cause. There are several variations between full-time and contract employees that must be recognized in order to maintain a happy working environment. To please each of these distinct classes, you must first recognize their differences and then use that knowledge to answer their specific concerns.

Employees on a Contingent Basis vs. Permanent Employees

- They are two different staff categories. The language used best describes the key difference between permanent and temporary employees. The latter group is employed on a contractual basis, and their work is dependent on the continuity of a particular situation. A temporary worker is often expected to adapt to surges in demand, but a full-time hiring is impractical because the demand would wane.
- Each section is valuable. A dependent worker is technically unique since they are handled individually by tax and job law. One should deem their permanent employees to be a desirable business commodity, and temporary staff can also be helpful. They may be retained to cover demand shortages, minimizing the expense of recruiting more workers when the need is only temporary.
- They have varying reasons. These people will be an important complement to the staff if they can take up any of the more repetitive duties that take up the usual personnel's time. Your full-time workers work for various reasons, including wages, benefits, and job satisfaction. In terms of incentive, a contract worker is equivalent to a full-time employee, but there are additional objectives. Many temporary employees are searching for ways to progress their jobs or have personal motives for working on a contract basis.
- Keep your long-term workers happy. Since these workers are the lifeblood of the business, you must build an experience that is empowering, satisfying, and inspiring. Contrary to popular belief, this would not necessarily imply an increase in pay. Increase employee rewards, discover special incentives, launch an employee appreciation scheme, or plan retreats that will keep them involved. Most notably, be ready to listen to and answer questions at all times. The morale of your permanent staff is primarily determined by their sense of being seen.
- Consider the particular conditions of your contract employees. A temporary worker is recruited to fill a vacancy in the workforce, and they are aware that their job could be terminated at any moment. Furthermore, their wages and all other benefits are predetermined, but they are not incentivized in the same way as your full-time employees are. When working with these people, you can't afford to be reckless or apathetic. As mentioned, they're a valuable asset to your company and deserve to be treated as such.

5.6 RECRUITMENT OF CONTINGENT WORKERS

Recruiting contingent staff provides the company with stability and versatility, helping it to respond quickly to change in an ever-changing world. Contingent staff may help companies fill talent holes, satisfy seasonal needs, and also acquire professional experience only when they need it. When you get a job order for a contingent worker, the recruiting process can vary from when you're looking for permanent workers. Be sure you completely appreciate what your customer needs: whether you're contracting or supplying contract staff. You and your client must be clear in the job description that the position is not permanent. Reiterate that this is a temporary place. Also, be meticulous. And if they are not a definite addition, the contingent worker should also suit the client's work culture.

5.6.1 FLEXIBILITY

When the company grows and changes, relying on contract labor or a "variable workforce" can be a good way to get through the ups and downs.

- Contingent employees insulate the company from market variations during the year, particularly for businesses who see seasonality or project-intensive efforts as a source of uncertainty in their workforce needs.
- Contingent labor reduces HR burdens and regulatory risk involved with recruiting and laying off employees as a result of downsizing.
- Using staffing services expands candidate outreach and reduces worker ramp-up period.
- Contractor off-boarding should be well-defined, fast, and simple.

By incorporating a more agile workforce and implementing a more just-in-time approach to hiring, you might be able to view market growth prospects with less risk.

5.6.2 COST

Any day of the week, one might argue that only a small number of firms actually recognize the true expense of a "fully loaded" employee. Can we know how to measure the costs of training, on boarding, incentives, overhead allocation, supplies, and so on? Should consultants get a higher price tag? If they are, I would suggest that the price isn't as high as you would expect, and that there are several advantages to consider:

- Sourcing, interviewing, hiring, disciplining, and firing temporary employees should be covered by staffing agencies.
- The individual is not paying federal taxes such as SUTA, FUTA, Social Security, or Medicare, nor are they bearing other insurance costs such as employer's compensation.

5.6.3 TIMELINESS

Speed and versatility go hand in hand, and in today's business world, speed is key. There are potentially crucial elements for fulfilling personnel timeliness demands:

- Temporary staffing companies will also deliver services quickly; this is crucial when a pressing need occurs, particularly if it is unanticipated.
- Staffing agencies will scan their database of registered jobs for suitable applicants and arrange for them to be shipped to the job site on short notice. These services would be in the pipeline of a true collaborator, with the requisite pre-employment activities completed.

5.6.4 SPECIALIZATION

Wouldn't it be wonderful to have the right squad of professionals on hand when you need them, Sort of like the old "Life Line" from that game show? Just use your talent when it's absolutely necessary. Recruiting firms have been extremely professional, with rich employee pools with deep expertise, allowing you to recruit candidates with specialized skills on an as-needed basis, typically on short notice. Accountants, attorneys, physicians, nurses, and information management practitioners are also included in several bigger companies that were once formed to serve administrative/clerical positions.

5.6.5 RECRUITING OPPORTUNITY

Contingent staff in your business should have an alternative pool of applicants with the following advantages:

- Contract employees are often a critical resource from which HRs may recruit for permanent roles.
- Many businesses use temps to fill in the holes when searching for full-time workers with the same skill set; if the contractor isn't the best choice, getting their input on the work can provide a better viewpoint to fine-tune the job description.
- In the very least, the temporary employee will do any or the entire job before HR considers the best long-term solution.

5.7 H-1B WORKERS

The H-1B visa was designed to allow qualified workers with a bachelor's degree or higher to work in fields such as computer science, architecture, pharmacy, dentistry, engineering, accounting, and more to work in the United States. Employers who want to file H-1B cap-subject petitions for fiscal years 2021–2022, including advanced degree exemption petitions, must first register online and pay the $10 H-1B registration fee. Each H-1B employee should only have one entry submitted by each employer. If their application is chosen, the employer must certify that they will submit a full H-1B petition.

Employers or their approved members requesting H-1B jobs subject to the cap can complete a verification process that includes only detailed details about their business and each requested worker. The eligibility dates for the United States Citizenship and Immigration Services (USCIS) is March 9–25, 2021. In case of any technical registration or site problems, it's wise not to wait for the last day of the lottery duration to request your entry. Those electronic registrations would then be subjected to the H-1B random selection process. H-1B cap-subject petitions may be open only to people with specific registrations (Figure 5.2).

The daily cap lottery is performed first, followed by the master's cap lottery in the United States. Individuals with a master's degree from the United States have an even greater risk of being chosen according to the drawing sequence. In fact, as

FIGURE 5.2 H-1B workers.

compared to the old ordering system, USCIS reports that it increases the chances of advanced-degree holders being picked by up to 16%. The USCIS issues H-1B visa to people who work in specialized professions that require a four-year bachelor's degree or similar experience. Many businesses rely on H-1B jobs to fill vital market needs or to acquire critical technological expertise and experience that aren't readily available in the United States. Where there are temporary shortages of required expertise, H-1B staff can be used. Employers also require H-1B staff to have advanced experience of international markets or to work on programs that help US firms succeed internationally. Employers would pay H-1B employees the median wage for the jobs they are doing, which is a crucial condition. As an H-1B employee, you can work for a US employer for a span of six years in a row.

H-1B visa holders may qualify for permanent residency with the aid of their employers. Their H-1B visas may be extended in one-year extensions during the processing periods before their green card is issued. When a foreign worker's H-1B visa expires, he or she must leave the United States for one year before filing a new H-1B petition. India accounted for 58% of all approved H-1B petitions in 2011, led by China (9%), Canada (4%), the Philippines (3%), and South Korea (3%). The US Congress imposes an annual limit on the number of H-1B visas that may be awarded per year, but the number of visas currently issued always exceeds the cap. The limit has been set at 65,000 visas since 2004, with an extra 20,000 visas reserved for overseas graduates with advanced degrees from US universities. Only those IT specialists, such as programmers and engineers at private technology firms, are subject to the cap. Scientists working as teachers at American universities, working in government research laboratories, or working with nonprofit groups are among those who are excluded from the limit. IT-related occupations accounted for about half of all H-1B recipients in 2012. Companies should bear in mind that even highly qualified and seasoned H-1B employees require assistance with their English skills while eval-uating the use of H-1B visa workers.

5.7.1 H-1B VISA APPLICATION PROCESS

The majority of businesses make ethical recruiting choices depending on how well an employee meets the work requirements. Since choosing to recruit the best pos-sible applicant, such businesses consider the need for an H-1B visa. To be eligible for an H-1B visa, a person must have a work offer from an organization that is willing to sponsor them. An employer must initiate the application process after deciding to recruit a worker who would need an H-1B visa.

There are two steps of the application process:

- The Labor Condition Application (LCA) and
- The H-1B visa application.

The organization submits an LCA to the DOL, specifying the job title, the geo-graphic region where the worker is required, and the estimated wage. The Wage and Hour Division of the DOL examines the LCA to ensure that the international work-er's earnings do not fall behind those of an American worker. Following certification of the LCA, the employer can apply to the USCIS for an H-1B visa, specifying who will fill the role and describing the person's skills and qualifications for the position. The USCIS must process the application before the applicant can be recruited, which can take several days or months.

H-1B is a nonimmigrant, employment-based visa that is issued to a qualified tem-porary worker. It is also known as a US job visa. Your company must file an H-1B visa appeal with the US Immigration Department, and you must have a specialized career in the United States. Once approved, this petition becomes an I-797 work permit, allowing you to acquire a visa stamp and begin working in the United States for that employer.

H-1B eligibility: An H-1B visa is issued for a specialization occupation, such as science, technology, engineering, arts, or mathematics, that requires the theoretical and practical application of a body of specialized knowledge and requires the visa holder to have a bachelor's degree or its equivalent. Job authority for H-1B visas is exclusive to the sponsoring employer. For more information, see H-1B eligibility and H1 petition process. Job authority for H-1B visas is exclusive to the sponsoring employer. H-1B visa stamping and interview: If your H-1B petition is accepted and you have a good visa interview with a US consulate or embassy, you can obtain an H1 visa stamp in your passport (Figure 5.3).

If you're applying for an H-1B visa from India, the procedure is as follows:

Step 1: Take a digital picture of yourself.

A digital photograph is required for each H-1B visa application. For certain visa types, photographic photographs are required, and for others, pictures are required. The approval of your digital picture or photo is depen-dent on the embassy or consulate where you apply in the United States.

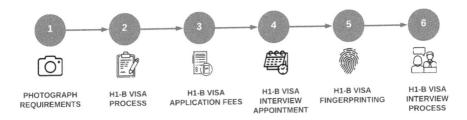

FIGURE 5.3 H-1B visa-tampering process.

Step 2: Fill out the DS160 H-1B visa application form.

The H-1B visa application (form DS160) must be filled out completely online. The application can be completed on the Department of State's website. You can get a confirmation with a 10-digit barcode after successfully completing the application. This page should be printed. This would be required in order to arrange your visa interview.

Step 3: Pay the processing fee for an H-1B visa.

The fee may be charged in cash or electronically through a bank transfer at specified Citibank or Axis bank branches throughout India. Build a profile on the US Visa Service website and pick the Plan Appointment option to ensure the correct amount is charged and triggered in a timely manner. On the payment approval page, you'll see payment options and more specifics about how to make a payment. The invoice is valid for a year after it is paid. For your visa consultation, you must make an appointment within a year.

Step 4: Arrange a visa interview appointment.

Following payment of the payments, you must make two appointments as part of the H-1B visa application process. The two-part appointment process begins with a visit to one of the five offsite Facilitation Centers to provide Biometrics detail, such as fingerprinting and photographing, and ends with a personal visa interview at the Consulate or Embassy.

Step 5: US Visa Fingerprinting at Visa Application Center (VAC).

Since scheduling a personal interview with the consulate or embassy, make a fingerprinting appointment at least a day or two before the consulate interview. Fingerprints were taken earlier during the personal interview. However, under the current rules, you must have biometrics details at your VAC appointment. This measure was adopted in order to ease congestion at US consular facilities and expedite the application process. Prior to the VAC appointment, arrange a visa consultation at the Consulate or Embassy.

Step 6: Attend a visa interview at a US consulate of your choosing.

Go to the consulate where your interview is held on the day and time stated in your interview appointment. Bring all needed and supporting records, as well as originals.

5.7.2 Potential Exploitation of H-1B Process

Despite the fact that firms applying for H-1B visas would pay at least 95% of the average wage for the profession, some businesses use H-1B visas to save costs?

Unethical businesses will get around the overall pay rule because salaries in the IT sector vary tremendously. At best, calculating a fair wage is an imprecise science. For example, an H-1B worker could be categorized as an entry-level IT employee when occupying a job that would cost $10,000–$30,000 and more per year if filled by an experienced worker. As demonstrated by the complaints brought against Vision Systems Group, dishonest firms may find other ways to get around the H-1B program's salary security. Companies, who employ H-1B employees, as well as the workers themselves, must think about what will happen when the six-year H-1B visa period expires. The visa program's transient existence can be daunting for both the sponsors and applicants. If a worker is refused a green card, the organization will lose a worker without developing a long-term employee. Many of these migrant workers are forced to uproot their families and return home after being unexpectedly unemployed.

5.7.2.1 Benefits of H-1B Visa for International Employees

The recipient, or foreign worker, who receives the visa, receives the most benefits. The USCIS received over 236,000 H-1B visa applications in 2016. A worker in a speciality occupation is permitted to be paid by an employer. Furthermore, because immigration law allows one immigrant employee to be the beneficiary of many H-1B petitions, H-1B beneficiaries can collect money from multiple employers.

5.7.2.2 Requirements in General

The first H-1B visa gain, and perhaps the most significant reason for its success, is the visa's broad eligibility criteria. Although certain other work visas require a management position, a master's degree, or a substantial financial commitment, the H-1B visa only requires the following:

- A bachelor's degree.
- A career offer from a corporation in the United States for a specialized work.

Of course, locating an employer with a job vacancy for a career similar to your educational experience is not easy, but these criteria are much easier to meet than the requirements for visa classifications such as E-2, L-1, and O.

5.7.2.3 Residency Period

Another advantage of the H-1B visa is the initial period of time that you are given before you obtain the visa. In comparison to other visas such as the B-1, which only allows you to remain for six months, and the J-1, which only allows you to stay for one year, the H-1B allows you to stay for three years and can easily be extended.

5.7.2.4 Adaptability

Unlike certain other work-related visas, the H-1B allows its holders to transfer their status from one employer to another while they are in the country. They will also work part-time and with several firms at the same time.

Anyone is qualified to apply.

One of the most appealing aspects of the H-1B program is that it allows international experts from all around the world to apply. The E-2 visa is only available to citizens of treaty countries, and the TN is only available to Canadians and Mexicans, but the H-1B visa is applicable to nationals and citizens of any country.

5.7.2.5 Dependents and Partners

Fortunately, the H-4 visa allows you to carry your partner and children over as dependents. Their stay will be limited to the length of the principal holder's, with extensions available for each H-4. Additionally, after receiving Job Authorization Papers, H-4 visa holders can work in the United States.

5.7.2.6 Dual Purpose

One of the H-1B's main advantages is that it is considered a "dual purpose" visa, unlike many other work visa classifications. This ensures that when in H-1B nonimmigrant status, you can seek lawful permanent residency. This is a major advantage over other visa classifications such as the TN and J-1.

5.7.2.7 Employers in the United States Benefit from the H-1B Visa

International professionals aren't the only ones who would benefit from H-1B visas. Employers, too, have a lot to win. This visa, which is for people with degrees in science, technology, engineering, or mathematics (or STEM for short), helps the best professionals from around the world to come to the United States and work for the business.

5.7.2.8 Local Labor Shortages Must Be Tackled

Many employers are based in locations where there is a labor shortage, and they may not be able to find skilled Americans willing to migrate permanently to your city. Employers may use the H-1B process to recruit temporary professionals from other countries to complete programs or fill specialist vacancies.

5.7.2.9 Competitiveness in the Global Market

You need to have a multinational footprint these days to thrive in today's economy. If you're able to support candidates for employment-based green cards, recruiting H-1B employees will improve the foreign competitiveness and have better permanent jobs.

5.8 WHISTLE BLOWING

When a worker discloses facts about corruption, it is referred to as whistle blowing. This is referred to as "making a disclosure" or "blowing the whistle" in this guidance. Most of the time (but not always), the corruption was something they saw at work. A worker who makes a leak must reasonably assume two items in order to be shielded by whistle-blowing legislation. The first is that they are behaving in the best interests of the group. This ensures that personal concerns are generally not covered by whistle-blowing laws. The second rational assumption that a worker may have is that the disclosure appears to indicate actual, present, or probable potential misconduct in one or more of the following categories:

- Illegal violations (financial impropriety, such as fraud, is one example)
- Failure to comply with a lawful duty
- Miscarriages of justice
- Endangering someone's health and welfare
- Environmental harm
- Covering up misconduct throughout the aforementioned categories The Employment Protection Act of 1996 includes rules on whistle blowing (as amended by the Public Interest Disclosure Act 1998).

If a person has been abused at work or has lost his or her career as a result of "blowing the whistle," he or she has the right to take his or her complaint to an employment tribunal.

5.8.1 EMPLOYERS RESPONSIBILITIES TO WHISTLE BLOWING

The best practice for an employer is to create an inclusive, clear, and healthy working atmosphere where staff feel secure speaking up. While the legislation would not mandate companies to have a whistle blowing program in effect, having it demonstrates an employer's willingness to listen to employees' complaints. Through establishing transparent rules and practices for coping with whistleblowers, an organization shows that it values knowledge brought to management's notice (Figure 5.4).

Staff are often the first to recognize some form of fraud within a corporation. Workers' findings can deter corruption, which could harm an organization's credibility and/or results, as well as save people from harm or death.

FIGURE 5.4 Employer's responsibilities with regard to whistle blowing.

When an organization does not have a transparent and inclusive atmosphere, a worker will be reluctant to make a report for fear of repercussions. The fear of retribution as a result of making a leak, as well as the assumption that no action can be taken if they do wish to "blow the whistle," are the two key obstacles that whistleblowers encounter. Information gathered by the Mid-Staffordshire NHS Foundation Trust Public Inquiry, the Freedom to Speak Up Independent Review into building an inclusive and truthful atmosphere in the NHS2, and the Parliamentary Commission on Banking Standards have both shown that many employees are unable to speak up about poor practice. The most critical move in developing an inclusive culture is ensuring that your workers can contact managers with serious questions. Employers should prove that they welcome and promote staff to make disclosures by showing clear leadership at all levels of the company.

Training and encouragement: To ensure that employees can effectively approach a variety of individuals in the organization, an organization can incorporate teaching, mentoring, advice, and other support programs. Dealing with a whistle-blowing disclosure when it is first posed by a worker is in the organization's best interests. This enables the agency to investigate quickly, ask a worker more questions, and, if appropriate, provide input. The advantages of making a disclosure should be clarified in a policy. Managers provide more intelligence to make decisions and manage risk in organizations that accept whistle blowing as an essential source of knowledge. When whistleblowers think they are being heard, they are more likely to come forward. If a worker may make a disclosure directly rather than going to a third party, the company gains. This way, there is a chance to move effectively on the details and correct any misconduct that is uncovered.

5.8.2 WHISTLE-BLOWING PROCESS

1. Determine the severity of the situation

    An individual should have clear evidence that his or her business or a coworker is behaving unethically and that the behavior poses a direct danger to the public interest before contemplating whistle blowing. The employee should seek out reliable outside resources and inquire for their opinion in a private and casual way. Do they consider the case to be critical as well? Their point of view can assist the employee in seeing the situation in a new light and ease concerns. Outside tools, on the other hand, could confirm the employee's initial concerns, resulting in a series of challenging ethical decisions.

2. Begin the documentation process

    When an employee notices a possible immoral or unethical activity, he or she should begin collecting ample evidence to prove misconduct. Incidents and facts, as well as the employee's perceptions of the situation, should be recorded. If legal evidence is required in the future, this document will assist in the creation of a chronology of events. Before moving on to the next stage, an employee can locate and copy all supporting memos, emails, manuals, and other papers. Records will vanish and become unavailable if this does not happen. Throughout the operation, the employee should retain paperwork and keep it up to date.

3. Making an effort to fix the situation internally

After that, an individual should try to resolve the issue internally by submitting a written summary to the relevant supervisors. Ideally, the employee would be able to identify the issue and address it from inside the company. The emphasis should be on revealing the truth as well as how the case impacts others. The employee's main aim should be to address the dilemma rather than delegate responsibility. This move should not be ignored or taken lightly, considering the possible negative effects of whistle blowing on an employee's future. Fortunately, several conflicts are settled at this stage, and the employee's more dramatic measures are needless. The relevant supervisors are approached and the situation that caused the whistle blower's intervention is addressed. Managers who indulge in immoral or criminal conduct, on the other hand, cannot accept an employee's questions or complaints. Whistleblowers should hope to be actively discouraged from taking any steps in such situations. Employees can be demoted or dismissed as a result of misleading or exaggerated statements. Attempts to smear the employee are very possible.

4. Consider taking the situation to the next level inside the company

It's likely that an employee's first effort to cope with a problem internally would fail. At this point, the employee can rationalize by raising the problem that he or she has completed all of the necessary tasks. Others may feel pressured to take further action as a result of their positive feelings about the case. As a result, a committed and attentive employee can be compelled to choose between worsening the issue and heading over the manager's head, or coping with the problem beyond the company. The employee should feel pressured to speak up.

5. Consider the implications of being a whistleblower

When whistleblowers think they have made a good effort to address the issue internally without success, they must take a step back to consider whether they are willing to move ahead and blow the company's whistle. Depending on the circumstances, an employee may be forced to pay significant legal fees in order to file or pursue claims against a company or entity that has access to a wide range of legal services as well as significantly more money than the employee. If an employee wishes to continue, he or she may be accused of holding a grudge against the company or of attempting to benefit from the allegations. The employee will be sacked, and his or her employers, colleagues, and even family members may lose faith in him or her. Before determining how to proceed, a possible whistleblower must try to answer a variety of legal concerns.

6. Develop an action plan with the help of experienced resources

A whistleblower should seek legal advice from a qualified attorney with expertise in whistle-blowing litigation. Based on the organization, employer, and state involved, as well as the type of the situation, he or she will decide which statutes and laws apply. Counsel should also be mindful of the statute of limitations for reporting the incident, as well as the law's protection for whistleblowers. Before publicly blowing the whistle, the employee should have an accurate appraisal of the strength of his or her legal position, as well as an expense estimate for a case.

7. Put the action plan into action

A whistleblower who wishes to take a lawsuit to court can do so by undertaking analysis and working with legal counsel. The best course of action for a whistleblower who wishes to stay confidential is to leak information to the press anonymously. However, the issue with this approach is that anonymous allegations are frequently ignored. Acting closely with relevant enforcement bodies and legal entities is most likely to yield positive consequences, such as fines, activities stopping, or other acts that draw the offending organization's immediate attention.

8. Consequences must be accepted

Whistleblowers must be wary of retribution, such as being discredited by colleagues, being harassed, or being set up; for example, management might try to move, demote, or fire the whistleblower for violating a small law, such as arriving late to work or leaving early. Management can argue that such behavior has been going on for a long time in order to justify their acts. To combat such acts and pursue legal redress, the whistleblower should have a clear plan and a good solicitor.

5.9   OUTSOURCING

Another way to satisfy manpower demands is to outsource. Outsourcing is a long-term outsourcing arrangement in which a corporation hires an outside company to perform a certain role. A business can employ an agency to provide services such as data center administration, telephone network support, or computer help desk personnel. Outsourcing has little legal conflicts of co-employment since the organization that pays for the services normally does not supervise or monitor the contractor's workers. Outsourcing is mainly used to minimize costs, but it is also used to achieve business stability to retain workers focused on the company's core competencies. IT executives launched the outsourcing movement in the 1970s, when they began supplementing their IT workers with contractors and consultants. This practice gradually led to businesses outsourcing whole IT business divisions to companies such as Accenture, Electronic Data Systems, and IBM, who could run a company's data center and execute other IT functions (Figure 5.5).

- Offshore outsourcing

Offshore outsourcing is a form of outsourcing where services are delivered by a business whose workers are based in another nation. Not just IT job, but any work performed at a reasonably high cost in the United States may be a candidate for offshore outsourcing. IT experts, on the other hand, will do a lot of their work from anywhere, whether it's on the company's premises or thousands of miles away in another world. Furthermore, by reducing labor costs by offshore outsourcing, corporations can save a lot of money. As a result, and since many foreign countries have a vast supply of skilled IT specialists, offshore outsourcing in the IT sector is popular.

FIGURE 5.5   Outsourcing.

According to a 2010 study, 93% of multinational businesses have engaged in some form of IT outsourcing. Twenty-three large corporations such as American Express, Aetna, Compaq, General Electric, IBM, Microsoft, Motorola, Shell, Sprint, and 3M use offshore outsourcing for help-desk service, network administration, and information technology growth. If more companies switch their core operations overseas, IT service providers in the United States are being pushed to drop their rates. Many American tech companies have established production centers in low-cost foreign countries where they can access a large pool of well-trained workers. Intuit, the manufacturer of Quicken tax apps, currently has offices in Canada, the United Kingdom, and India. Wide production centers are based in India for Accenture, IBM, and Microsoft. Cognizant Technology Solutions is based in Teaneck, New Jersey, but the company's technology centers are mostly in India. Because of the high wages paid to technology developers in the United States and the simplicity with which clients and vendors can collaborate, large programming ventures are increasingly being outsourced overseas.

5.10   INFORMATION AND COMMUNICATIONS TECHNOLOGY CODE OF CONDUCT

1. Information and Communications Technology (ICT) Community performs business in a fair and honest manner in line with best business practices.
2. ICT complies with the laws and regulations of the countries in which it operates, including relevant legal standards for employee, third-party, and environmental health and safety.
3. ICT employs and tests people purely on the basis of their skills and credentials for the role they are filled.
4. ICT aims to create a healthy, open culture free of bigotry, coercion, and/or bullying, in which all workers have fair access to information and resources.

ICT is dedicated to fostering a dynamic society that values free and fair dialogue and places no restrictions on what is allowable.

5. ICT may not allow staff who are deployed at clients to offer goods and/or services that the clients have not ordered, or to participate in some other form of conflict of interest.

6. Corporate social responsibility is rooted in the company's ethos. ICT aspires to operate with honesty and accountability, and to be accountable to its owners, staff, community, and society.

7. The essence of the underlying transactions must be correctly defined and represented in ICT's administrative books and records and related documentation. There should be no confidential or volatile deposits, portfolios, or assets opened, obtained, or retained.

8. ICT employees' personal financial activities or interests must not conflict with ICT's activities or interests. Conflicts of interest must be reported immediately.

9. If ICT workers have material, nonpublic knowledge that counts as prescience, they will:

   a. If the nonpublic information is relevant to ICT and they are held to the legal guidelines on prescience as well as the internal ICT insider rules, they should refrain from engaging in transactions in ICT Group NV shares (including the exercise of an option).

   b. Refrain from engaging in transactions in ICT Group NV shares (including the exercise of an option) if it is the nonpublic information.

10. Variables other than business concerns do not affect ICT workers' business decisions. Third parties doing business with ICT do not assume that offering business gifts and/or other favors to particular workers would help them. Bribery, in any manner, is unacceptably solicited and approved, and will result in instant dismissal.

11. ICT staff and vendors may not give payments or gifts to (employees of) companies or third parties, or make explicit promises, in order to secure company contracts or to favorably affect a transaction or acquisition process. This ban doesn't include corporate gifts not exceeding EUR 75.00.

12. ICT is forbidden from providing donations to political parties, associations, or their members. Participation in the society is welcomed.

13. ICT workers will adhere to national and international antitrust laws and refrain from entering into arrangements with other businesses that (may) limit competition, such as costs, distribution terms, market sharing, output, and marketing.

REVIEW QUESTIONS

1. What are social networks and how do they work? Discuss how social media can be used for business.

2. Discuss the most important legal questions around the usage of social networking websites.

3. What does it say to blow a whistle? What constitutes an effective whistle-blowing procedure?
4. Elaborate the process of H-1B Visa.
5. What are the ethical issues involved in whistle blowing and protection of whistle?
6. How an organization decides to use contingent workers? What are the advantages and disadvantages of contingent workers?

MULTIPLE CHOICE QUESTIONS (MCQs)

1. **What form of contingent worker is more likely to acquire advanced skills that are in short supply in the job market?**
   a. part-time employees
   b. temporary employees
   c. leased employees
   d. independent contractor
2. **Contingent workers**
   a. share jobs with other employees
   b. are only hired when they are needed
   c. always work less than 40 hours per week
   d. around the same moment, function with many employers
3. **Most people who are categorized as nontraditional workers are ____.**
   a. contingent workers
   b. independent contractors
   c. job sharing
   d. working multiple jobs
4. **Which of the following is a sort of nontraditional worker?**
   a. contingent workers
   b. part-time workers
   c. people working in alternative work arrangements
   d. people with multiple jobs
   Select the answer from the codes given below:
   a. a, b, & c
   b. a, b, & d
   c. b, c, & d
   d. a, b, c, & d
5. **Which of these activities are not included in the scope of human resource management?**
   a. Job analysis and design
   b. Motivation and communication
   c. Safety and health
   d. Organizational structure and design
6. **The downside of telecommuting is that workers at home are likely to**
   a. work on weekends
   b. face distractions at home
   c. feel isolation from coworkers
   d. feel exploited

7. **Offshoring is**
   a. substituting foreign for domestic labor
   b. subcontracting a portion of the supply to a certain business
   c. exporting
   d. importing

8. **Outsourcing is**
   a. exporting
   b. importing
   c. a firm having someone else does part of what it previously did itself
   d. constructing a plant in a different nation to cater to that country's demand

9. **Which of the following is an implied duty on the part of an employee?**
   a. To commit a gross misconduct offence
   b. To organize holiday entitlement
   c. To be ready and willing to work
   d. To pay Salary

10. **Which of the following best describes a "restraint of trade" clause?**
    a. An employee's freedom to get another career after they quit their current job is covered by this provision.
    b. An employer's attempt to prohibit workers from quitting their jobs by the use of a provision.
    c. A provision that forbids a company from providing the national minimum wage to its workers.
    d. An employer's attempt to defend their corporate assets if an employee exits their role.

ANSWERS TO MCQs

Q1: d, Q2: b, Q3: b, Q4: d, Q5: d, Q6: c, Q7: a, Q8: c, Q9: c, Q10: d

6 Impact of IT on Quality of Life

LEARNING OBJECTIVES

- To discuss recent technological advances on quality of life
- The use of telephone, computer, and Web/Internet-based technologies
- To bridge the gap between society's poor and the wealthier, middle-class residents of cities and suburbs
- To study the advancements in health care systems

6.1 INTRODUCTION TO IMPACT OF IT ON QUALITY OF LIFE

The standard of living differs considerably between classes within a society and from country to country. Gross domestic product (GDP) per capita is the most commonly used metric of a person's material quality of living. The cumulative annual production of a country's economy is referred to as national GDP. In general, developed nations have a higher standard of living than emerging countries. The quality of living in the United States, as in other developing nations, has increased over time. However, economic cycles influence costs, incomes, job levels, and the output of products and services, so the rate of change varies. Significant disasters, such as earthquakes, hurricanes, tsunamis, and war, will wreak havoc on people's lives. The worst economic decline in American history happened during the Great Depression, when GDP plummeted by around 50% between 1929 and 1932, and the unemployment rate hit 25% by 1932. Investment and productivity refer to the amount of output produced per unit of input, and it may be measured in a number of ways. For example, at a factory, productivity is measured by the number of labor hours required to manufacture one item, whereas in the service sector, productivity is determined by the yearly sales produced divided by the annual pay received.

Most countries have been able to increase their output and services over time—not by increasing input proportionally, but by improving output efficiency. Since the average hour of labor produces more goods and services, these productivity gains have resulted in an improvement in the GDP-based standard of living. The Bureau of Labor Statistics tests productivity in the United States on a quarterly basis. Over the past century, labor productivity growth in the United States has been about 2% every year, indicating that living standards have increased twice every 36 years. Between 1947 and 1973, productivity grew at a rate of 2.8% each year, due to new organizational methods and automated technologies that rendered employees even more productive (Figure 6.1).

DOI: 10.1201/9781003280989-6

FIGURE 6.1   Importance of technology.

Productivity fell in the mid-1970s, then rose in the early twenty-first century before plummeting again from 2007 to 2012, coinciding with the United States' worst recession since the Great Depression. IT has played an important role in encouraging creativity and is a crucial factor in growing efficiency. IT, as well as other innovative technologies and financial spending, was used by progressive management teams to introduce product, operation, and service advancements. Productivity increases were simple to calculate in the 1960s, when information technology (IT) was in its infancy. Midsized firms, for example, often had a dozen or so accountants devoted exclusively to accounting for payroll. Businesses required fewer accounting workers as they introduced electronic payroll processes. The benefits of such IT expenditures in terms of production are evident.

- Additional variables that can influence national production rates are summarized in the following list:
    - The rate of increase in labor productivity varies depending on whether a country is in an expansion or contraction phase of the economic cycle. Firms should take full advantage of economies of scale and full output during growth periods. When the economy is in a recession, there are less investment opportunities.

- Outsourcing will distort efficiency if contracting companies and outsourcing firms have different productivity rates.
- In comparison to many other developed countries, regulations make it possible for businesses in the United States to recruit and dismiss employees, as well as start and stop company operations. Markets can more quickly move employees to more profitable businesses and industries as a result of this versatility.
- To stay ahead of the competition, more open markets for products and services will create greater incentives for technical advancement and acceptance.
- In today's service-based economy, calculating the actual performance of resources such as payroll, customer service, and consultancy is challenging.
- IT investments don't necessarily yield concrete returns such as cost savings and decreased headcount; instead, many of them produce intangible gains such as increased efficiency, reliability, and service.

6.2 IMPACT OF IT ON THE STANDARD OF LIVING AND PRODUCTIVITY

When we think about "standard of life," we generally think of whether or not we are rich. Simple needs such as sufficient food, clothes, and shelter are necessary for happiness. Since it is an economic term commonly calculated by standards such as actual income per person and poverty rate, we clearly ask people, "How is your quality of living today?" There are also other factors that influence one's standard of living, such as health insurance, life expectancy, income inequalities, and educational expectations. People's levels of living are constantly growing as the economy grows, and these improvements are substantial over time.

In certain cases, we can assume that efficiency is the most significant determinant of living conditions. Regardless of whether we compare figures from the same world over time or from different countries over time, it is apparent that only a solid economic base will guarantee a high standard of life. Many people might have questions on which is the independent variable and which is the dependent variable in these two variables: living level and efficiency. The fourth industrial revolution's technological developments have culminated in monumental and unprecedented shifts in the world economy, culminating in modern globalizing systems that are highly collaborative and sophisticated. Instantaneous connectivity, modern products and industrial processes, better air and sea transportation, and a boom in the building industry have all been made possible by technology. Working from home using online channels and making restaurant orders and personal transportation services through the Internet have all allowed those with access to such technology to better their livelihoods and even live in comfort. In terms of health changes, cellphone technologies in developed countries will open several doors for health literacy, banking, service availability, and language translation. Cellphones with Internet connectivity are becoming more popular in those areas as they become affordable and more easily accessible, helping disadvantaged people to access online content and resources that they did not previously have access to.

There is certainly a connection between efficiency and living standards, but it is not a clear one. A precise mathematical formula exists in algebraic form to describe this relationship. The link between competitiveness and living standards has far-reaching implications for not just a country's governmental policies, but also a company's strategic plans. In the domains of macroeconomics and microeconomics, they are as follows: Raising productivity in a society may improve people's living circumstances and make the nation more peaceful; in a company, it can enhance labor efficiency and allow for more profit to be produced. Nature's goals are the same as theirs.

The government is exploring a number of alternatives for economic growth. According to a recent theory, the increasing contribution of productivity to economic growth is due to the relatively high cost of technology as well as its exporting through commerce. Domestic use, which is in high demand, is also a significant factor. Productivity gains will have a broader impact on society by increasing living conditions and generating profits. This, though, is the product of a collaborative endeavor spanning research, technology, education, and virtually every area of society.

It is a little better for corporations. Productivity can be improved in many ways. Of all processes, automation and computerization are the most apparent. Materials, labor, and machinery are the three main factors that make up production. The general rule is to decrease the amount of activities that workers must do, causing robots to become increasingly relevant and have a huge influence on workplace efficiency.

Laptops and computers are now aiding in the development of work in developed nations in a number of ways. Users who have access to the Internet may not only interact with employers and look for employment online, but they also act as a powerful platform for training, literacy, and education. Students who have access to computers will learn to read and write, as well as gain advanced skills for the future, through visiting websites or using applications such as Microsoft Word. Impact procurement, which is regarded as an innovative market-based approach to poverty alleviation and has the potential to generate millions of opportunities for those living in poverty, is one of the most potential technological applications in underdeveloped countries. When companies recruit vulnerable employees in developed countries to perform activities that include human contact but do not necessitate physical intervention, this is known as effect procurement. Effect procurement is now having a positive impact on developed economies, with wage rises varying from 40% to 200%, in some populations participating in this sector. This industry was worth $4.5 billion in 2010 and employed 144,000 people in developed countries, and it's just getting bigger.

6.3   DIGITAL DIVIDE

The digital divide is the gap between people who have access to modern information and communication technology and those who do not. There are three major phases that have an effect on global digital injustice. Digital disparity exists between urban and rural populations, as well as between social classes; countries that are less

FIGURE 6.2   Digital divide.

developed economically and countries that are more developed economically, as well as educated and uneducated people, are all represented (Figure 6.2).

Digitally splitting people who have access to a broadband network is possible. The divide is widened by computers with poor performance and limited Internet rates, and access to subscription-based content is restricted.

There are several different forms of digital divides that hinder our ability to access the Internet. The following are some of the most glaring cases of digital inequity:

1. Gender gap

    According to a 2013 survey, gender disparity on the Internet is particularly pronounced in developed countries. Although mobile networking is rapidly expanding, it is not spreading uniformly. Women appear to lag.

    In low-income nations, men are 90% more likely to possess a mobile phone than women. This equates to 184 million women who do not have access to a cellphone. In low and middle-income countries, even though 1.2 billion women have cellphones, they have no access to the Internet.

2. Social disparity

    People with common preferences shape relationships and social circles as a result of getting access to the Internet. Twitter and Facebook, for example, build online peer networks focused on shared interests.

    Internet use has affected social stratification more than ever before, and is visible in society between those who are linked to the Internet and those who are not. Nonlinked groups are marginalized because they do not profit from the Internet in the same way as related groups do.

3. Gap of universal coverage

    When it comes to using the Internet, people with physical disabilities are often at a disadvantage. They can possess the requisite skills, but they are unable to take advantage of the available hardware and software. Owing to a shortage of computer literacy skills, poor education rate, and insufficient broadband connectivity, certain areas of the world will remain cut off from the Internet and its immense potential.

In the United States, there is a digital divide:

A fifth of the population of the United States is lacking access to broadband. The majority of people cannot pay the high monthly data plan fees. In the United States, only 56% of households making less than $35,000 a year have access to the Internet at home, compared to 92% of households earning $75,000 or more. Students are harmed by the technological gap.

They fall behind in their tests and are unable to contend with their peers. They miss out on chances to improve their financial condition, perpetuating the cycle of poverty. Although most schools have replaced textbooks with low-cost computers and Chromebook laptops, schooling comes to a halt as students return home and have no Internet connection. According to Deloitte, when a citizen does not have access to the Internet, America loses over $130 million every day.

In Connote Valley, most households link to the Internet with a satellite dish, which is costly and only connects at a snail's pace. The majority of the residents who are not related lack the requisite digital literacy skills.

According to the Federal Communications Commission, 24.7 million Americans, mainly in remote areas, do not have access to broadband. According to Microsoft, only 2% of the population in Ferry County has access to broadband. There are important measures that, if correctly taken, will help bridge the digital gap in America.

6.4 FACTORS ATTRIBUTING TO DIGITAL DIVIDE

The digital gap is growing at an unprecedented rate, despite the fact that the number of Americans with access to computers and the Internet continues to increase year after year. On the one hand, previously connected groups of society, such as higher income, better-educated White and Asian Pacific Islander families, are adopting newer technologies faster and networking more. On the other hand, groups with historically lower rates of Internet and computer usage do not trail far behind. Unfortunately, the digital gap is widening along already strained ethnic and economic divisions, according to a report commissioned by the National Telecommunications and Information Administration (NTIA) titled Slipping through the Net: Describing the Digital Divide.

6.4.1 EDUCATION

The digital gap seems to be widening as income levels rise; families with greater levels of education are more likely to use technology and the Internet. People with a college education or higher degree are ten times more likely to have Internet access at work than those with just a high school diploma, according to a research. The gap in computer usage and Internet connection between individuals with the greatest and least education grew by 7.8% and 25%, respectively, according to a study conducted by the NTIA between 1997 and 1998.

6.4.2 INCOME

Unsurprisingly, and in direct proportion to education, household income levels have a major role in the growing disparity. According to the NTIA study, "the gap between the top and lowest income groups has grown by 29% in the past year" (NTIA Falling through the Net 99). Families with earnings over $75,000 are 20 times more likely than those with lower incomes to have home Internet connections, and those who live in the city or suburbs are ten times more likely to have a computer than those who live in rural regions. Because of their lower income levels, bad neighborhoods lack the facilities seen in affluent regions. Greater access to Internet networks is available in wealthier regions, making it more appealing for new companies to set up shop. As a consequence, poverty makes it less appealing for foreign companies to invest in less prosperous areas, widening the divide.

6.4.3 RACE

At the same time, the digital divide is growing along obvious racial lines. Between 1994 and 1998, the gap in machine usage between White and Black families grew by 39.2%, and between White and Hispanic households by 42.6%. Hispanic families are approximately half as likely to purchase computers as White ones. Surprisingly, the amount of computers in the classroom is affected by race. Schools with a higher minority population have fewer laptops than schools with a lower minority population. As one would expect, disparities across ethnic groups decrease at higher income levels but widen within families at lower income levels. In terms of Internet access, Black and Hispanic families are falling farther behind: White households had 37.6% greater access between 1997 and 1998. Hispanic families are nearly 2.5 times less likely to utilize the Internet than White ones. Furthermore, according to the NTIA study, racial disparities in Internet access exist regardless of wealth. A cultural study of Hispanic, African-American, and Asian-American cultures was conducted in order to find reasons for the gap other than income. Computers were regarded as a luxury rather than a need in Hispanic society; computing habits separated individuals and diverted time away from family activities. In African-American culture, it has been observed that African-Americans have historically had negative relationships with technological advancements. Asian-Americans, on the other hand, placed a larger value on education, which led to a greater acceptance of technological advancements.

6.4.4 OVERCOMING THE DIGITAL DIVIDE

Bridging the digital divide is a huge and complex topic that is deeply intertwined with topics of race, wealth, and poverty. When broken down into unique tasks that must be done, the difficulty is far from insurmountable. Aside from the apparent financial obstacles, the following suggestions would aid in closing the gap:

1. Make it more affordable
   In both rich and developing nations, affordability remains one of the most important barriers to Internet adoption. Because of the high prices, a substantial number of people are unable to use the Internet. Smartphone and notebook computers are costly, considering the fact that technology is a necessity. Taxes, patent fees, and energy are all factors that contribute

to high technology costs. To assist with this, low-income earners are provided with loans in purchasing advanced technologies. Tariff subsidies may be offered by governments to enable people to purchase these software instruments.

2. Giving users more control

To realize the Internet's full potential and impact on the world, we must utilize its resources. The bulk of Internet users are only vaguely familiar with some of its applications. Google, for example, assists users in locating knowledge that they otherwise would not have. An issue known as "participation inequality" widens the digital divide by preventing individuals from accessing it because they lack the required skills.

Since user data are used in decision-making, the information gathered may not be adequate for proper decision-making, resulting in bad decisions with potentially disastrous results. To prevent this, the general public must be educated about the benefits and significance of utilizing the Internet and its many instruments to promote economic and social growth. Furthermore, the public should be able to post their thoughts and all other related usage data on the Internet in order to help the government and other organizations make sound choices that represent the interests of the people.

3. Increase the usefulness of Web content

According to study, importance is the most significant obstacle to Internet access in developed countries. This is because most people are unable to locate information, web and smartphone apps or Internet resources in their native language. Furthermore, the majority of people in rural areas lack the necessary information to comprehend some online material.

Local content and applications in local languages that the local people can understand are needed to facilitate Internet use in those countries. Aside from that, concerns such as anonymity, trust, and data protection, which threaten to turn off many prospective customers, must be resolved by developing policy mechanisms that ensure Internet websites secure their users' data and online activities.

4. Infrastructure growth for the Internet

The Internet uses infrastructure to transmit data between two or more computers located in different areas of the world. Due to a lack of adequate Internet connectivity, as is the situation in many third-world countries, some parts of the globe have only sporadic or no Internet access at all.

After the advent of broadband Internet, which is both faster and more efficient than traditional dial-up connections, this issue has grown much more apparent, especially in rural regions. The high cost of the networks and technologies required to relay broadband Internet makes it uneconomical to deploy in distant areas. Fortunately, large-scale, cost-effective rural alternatives have been developed, including satellite broadband technologies, drones, and Earth-orbiting balloons.

5. Address the disparity of Internet connectivity between men and women

According to 2016 Internet consumption statistics, there are 250 million fewer women online than men. The bulk of these women are from

Africa and Arab countries. This suggests that, in order to bridge the digital gap, women must be given extra attention in order to get their consumption closer to or on par with men's. The fact that there are more women without cellphones than men in low and middle-income nations exacerbates the problem. The drive to bridge the gender divide falls heavily on the shoulders of both the government and nongovernmental organizations, which must collaborate and collaborate. The plight of women without Internet access can be significantly alleviated by addressing topics of poverty, importance, and public visibility.

6.5 UNIVERSAL ACCESS

The need for connection grows in tandem with the number of individuals who use computers and the Internet. The importance of such services must be understood by policymakers and community members in the public sector, and measures must be taken to guarantee that everyone has access to them. Despite increased competition among PC manufacturers and Internet Service Providers, many families still find the expenses of buying a computer and maintaining a home connection excessive. In the same manner that the government subsidizes basic phone service, the government should subsidize Internet connection for low-income households. Simultaneously, the private sector must commit to providing equitable infrastructure and networks to rural and underserved regions, ensuring that everyone has access to and benefits from them. Community Access Centers (CACs), with Existing Centers Continually Supporting Community Connection Centres, are a vital resource for individuals who do not have access to laptop computers or the Internet at school or at work; funding for these services should be maintained in order to develop and enhance them. According to data collected in 1998, minorities, those with lower incomes, those with less education, and the jobless—the groups most affected by the digital divide—are the primary users of CACs. CACs "use the Internet more often than other classes to locate employment or for educational reasons," according to the study (NTIA Falling through the Net 99). CACs, as a consequence, are clearly worthwhile expenditures.

Computers and other devices are insufficient without additional, well-trained technical staff. To encourage the best use of money, cities and schools must educate and maintain more skilled personnel, as well as introduce emerging technology. The workers must be able to educate others as well as understand the emerging technology.

Public's Outlook toward Technology Is Changing

At the same time, a segment of society must change its attitude toward technology. Instead of viewing computers and the Internet as luxuries, the general people may regard them as necessities. The general public must see new technology's enormous potential and embrace it as a source of opportunity for their own as well as their children's futures.

Programs Now Running

Given the vastness of the still-widening digital divide, any support is exceedingly helpful. Thankfully, the nation, nonprofit organizations, and private foundations have begun to introduce projects aimed at closing the gap. Although the following list of services and websites is by no means exhaustive, it does reflect a cross section of the numerous projects currently underway.

6.6   IMPACT OF IT ON HEALTHCARE COSTS

Continual scientific advances in healthcare have saved many lives and changed the lives of many more people. Not only has technology altered patient and family inter-actions, but it has also had a significant effect on medical processes and healthcare providers' practices. Electronic health records (EHRs) are often believed to mini-mize substantial and increasing healthcare costs while enhancing healthcare quality in the United States. However, the evidence for these connections is mixed, leading to doubts about EHR's efficacy. We contend that knowledge and patient exchange are the most important determinants of EHR's effect on healthcare costs and efficiency.

As healthcare investment in the United States tends to escalate at a rate that out-paces that of other segments of the economy, there seems to be a greater appetite and desire to look into the reasons of the surge and discuss policy alternatives. Some of the factors for increasing prices are related to an aging population that all too frequently takes bad personal decisions, but others are clearly linked to rises in the costs of medications, hospital and nursing home treatment, insurer insurance, and expensive medical equipment. Any healthcare economists conclude that the acceler-ated implementation of emerging technologies has also contributed greatly. Vendors naturally want to get a new device or invention into the market as soon as possible, but this may also happen before significant proof of profit is available.

Remote consultations with specialists, customized treatments, and the availability of user-friendly mobile phone software have also culminated in increased patient treatment and a stronger overall healthcare experience. Furthermore, patient's qual-ity of life has strengthened thanks to the availability of newer medical approaches that yield better outcomes.

6.6.1   DIGITIZATION IN HEALTH RECORDS

Patients and healthcare personnel can use digital archives that have been electroni-cally transferred to the cloud and made available online. Data collection, control, and as a consequence, transmission have become easy and quick. Professionals and practitioners would have access to support for treatment decisions, allowing them to make safer, more knowledgeable care decisions. The digitization of health data fur-ther increases quality and allows for the accessibility of healthcare to patients in rural or unavailable areas. This digitization has the potential to boost health outcomes while simultaneously lowering costs.

- Medical sector and smartphone app technology
  Patients may utilize software on their mobile devices to keep track of doctor's visits and remind them to take their medicines, in addition to pro-viding quick and accurate diagnostic data. Health and wellness apps help users stay healthy by keeping track of their food intake and activity levels, as well as offering individualized solutions.

These applications also support doctors in high-stress careers by cutting down on time spent reporting, maintaining records, and other mundane activities. Mobile apps include access to drug information to help prevent adverse effects and interactions, overcome problems, and enhance detection. Doctors can better coordinate with their patients, maintain track of their appointments and consultations, and increase the effectiveness of their treatments by carefully monitoring their vital signs.

- Health history in an electronic format

EHRs are computerized summaries of a patient's medical data. Information with regard to hospital visits, surgical procedures, and medicines, as well as diagnoses and test findings, may be included. They offer a detailed description of a patient's health, enabling for more accurate diagnosis and treatment.

These electronic gadgets allow labs and specialists to cooperate and exchange information without having to spend time and money on manual delivery. EHRs provide healthcare providers access to information regarding a patient's allergies and intolerances, as well as any other relevant data; this is particularly essential if the patient is unconscious.

EHR procedures, when appropriately managed and enforced, will also result in increasing transparency and reducing malpractice. Electronic archives are simpler to build and manage and require less time. They make medical accountants' work simpler and reduce the risk of errors. Although the value of digitization has been well recognized for some time, some companies and sectors are still reluctant to embrace it. One such business is healthcare, which is appropriate given the protection risks that technology faces.

Healthcare will now completely benefit from digitization thanks to advances in device and server security. It should be understood that healthcare is a trillion-dollar sector that includes patient care, clinical facilities, nursing homes, home healthcare, drugs, research and development, and that if they take advantage of those advances in combination with proper promotion of healthcare services, they will double or even triple the amount they get. Without a doubt, there is a lot of need for digitization in the healthcare industry. That isn't to suggest that there isn't any digitization of healthcare; automation is also helping to improve patient treatment. Although some healthcare practitioners are outspoken in their opposition to the emerging generation of technologies and healthcare, others who are quick to incorporate it into their practice have a distinct edge over their competitors.

- Coordination between the doctor and the patient is improved

Patients can keep track of their medical records more easily thanks to digitization. Gone are the days when any doctor's appointment required you to keep a paper file. When a physical copy of your medical records isn't present, you'll need it in an emergency. In this scenario, digitization aids the doctors by holding them up to date with the patient's medical records at all times. You can build and preserve a digital file in your medical records as a patient, upload it to a cloud server, and share it with the individuals

or physicians you choose. Of course, maintaining this requires some commitment as well as technological know-how, but it is well worth it in an
emergency.

- Administrative tasks may be automated

  Administrative costs account for more than a tenth of all healthcare costs
  in the United States. This is higher than any other nation in the first world.
  Doctors devote less than a portion of their time listening to patients, which
  is even more troubling. Instead, the majority of their time is "spent" on different administrative activities. Doctors will waste up to a third of their time
  simply inserting patient information into an EHR system.

  Clearly, this is an inefficient device, because a physician's time is a valuable asset. Here, digitization will shine and do what machines are notorious
  for: process automation. Advanced artificial intelligence (AI) programs that
  offer drug guidance and alerts may be applied. Administrative functions are
  clearly a major impediment to physician success, as shown by the statistics.
  Physicians and nursing professionals would have less work to do once these
  processes are automated, allowing them to concentrate on treating patients.

- Collaboration among multiple physicians

  Unfortunately, certain people have medical problems that necessitate
  consulting with a variety of specialists. It is important that the various
  physicians maintain a continuous line of contact in order to formulate the
  right overall treatment strategy for the patient. Conflicting prescriptions,
  for example, can cause problems, but with a mechanism in place that holds
  all clinicians informed, they can create a drug schedule that is free of dis-
  agreements. The experts will also know what is working and where change
  is being made thanks to continuous knowledge exchange. This isn't an
  unusual situation, and it's one where digitization will help doctors perform
  their jobs more easily and safely.

- Data protection

  Malicious attacks on digital records are still a possibility, but that doesn't
  mean physical files are entirely secure. There is still the risk of patient
  reports being lost due to unforeseen circumstances such as natural disasters (earthquakes, flooding) or neglect. These considerations can be elimi-
  nated from the calculation for medical history for a well-designed digital
  data-collection device. Not just that, however, approved employees may
  access these documents from anywhere, guaranteeing prompt availability
  in the event of a need. Appropriate fail-safe mechanisms can guarantee that
  the data remain secure and are not harmed by human error, natural hazards,
  or other factors.

- Health data in real-time

  Smartphone and wearables are very common these days, and they're
  becoming more capable all the time. Modern smart watches will monitor your heart rate, count your steps, and even do an ECG. This makes
  them very useful not only for regular patients but also for their doctors.
  Consider a patient who wears a smart watch that continuously transmits
  their health data to the Internet. The doctor will set a "trigger" for when a

certain condition is met (e.g., when a patient's heart rate exceeds a certain threshold), allowing them to respond quickly. In today's world, doctors will use real-time data analysis to respond proactively to prevent anything bad from happening.

6.7 BIG DATA AND CLOUD

In a number of sectors, including healthcare, big data has become a catchphrase. This is due to the ability to generate and collect massive amounts of data from a variety of sources in the healthcare industry. This information is subsequently utilized in analytics, which allows for the early identification of future outbreaks and, ultimately, the prevention of fatalities (Figure 6.3).

Data management in the cloud improves productivity and accessibility thereby minimizing waste. This aids research and development in the creation of new treatment regimens and life-saving drug compositions. Cloud services, in fact, may be very helpful to medical care, allowing for enormous amounts of research and study as well as efficient clinical data exchange. Without the hassle or price of maintaining additional server hardware, the cloud provides dependable and cost-effective storage choices, as well as backup and recovery capabilities. In the next decades, the healthcare industry, as well as the data connected with it, is projected to grow at an exponential rate. The data generated through EHRs, laboratories, medical equipment, and even patients themselves are expected to amount to petabytes, exabytes, or perhaps zettabytes. It would become more difficult for IT firms to analyze massive quantities of data and turn it into valuable medical data.

Big Data in healthcare, along with Cloud computing, will pave the way for new medical models. Earnings would certainly increase if market understanding is improved. The healthcare business can upload more data to the cloud, and Big Data

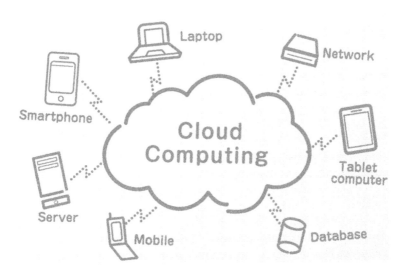

FIGURE 6.3   Importance of cloud computing.

analytics can extract insights from that data, paving the way for a more progressive future for the industry.

- Health data are being used more efficiently thanks to Big Data Analytics.
  Hospitals, retail and nonretail hospital operations, and advertising events create a vast volume of data on a daily basis. However, the bulk of it is squandered because the required individuals are unable to decide what to do with the details. This is where Big Data on the Cloud comes into play. Big Data modeling software and libraries simplify the task of generating accurate and calculative insights from vast numbers of data in seconds. As a consequence, more physicians trained in big data management will be required in the future. The Big Data revolution is bringing sophisticated methods of combining data from many sources to the forefront. The focus is on providing doctors and medical professionals with the most current and relevant information in real time as they interact with their patients.

6.7.1 Introduction to Cloud Computing

For large-scale storage and sophisticated analysis, as demanded by Big Data and Big Data Analytics, cloud computing is the most suitable architecture: Flexibility, security, parallel processing, scalability, and resource virtualization are just a few of the benefits. Cloud storage can lower automation, information, and infrastructure maintenance costs while also improving operating performance and user access. The drive for Big Data Analytics in healthcare, as well as the rising value of cloud computing, has brought healthcare new perspectives. Health data are increasingly being submitted to the cloud and shared by healthcare providers, and Big Data medical repositories are being used for analytics.

Every day, a vast amount of healthcare data is produced. The detail is crucial for making decisions and ensuring the best quality treatment to patients. Cloud storage is a cost-effective tool for capturing, storing, and sharing real-time data between healthcare organizations. Cloud computing provides a high latency and massive storage space, which are two important characteristics for effective data processing with large patient populations. One of the key issues about using cloud-based healthcare systems is security and privacy. To leverage the cloud technology, healthcare institutions must provide electronic medical records (EMRs). To keep up with the exponential advances of IT and the widespread use of cloud-based computing, attempts should be made to convert healthcare data from paper to electronic format. Then, to monitor and manage the use of healthcare records, regional laws and regulations should be implemented. The dramatic increase in average life expectancy has culminated in a population that is rapidly aging. This has culminated in an increase in the need for additional care and a wide range of medical treatment options. Creative and cost-effective strategies are required to assist healthcare providers and more efficient approaches to solving this global problem. Cloud computing offers functional applications, as shown by IBM and Active Health Management's new clinical information management system, "Collaborative Treatment Approach," which was launched in November 2010. The cloud-based technology was created with the goal of assisting medical and healthcare workers to rapidly access data and information from various sources, including EHRs. Patients with chronic illnesses, on the other hand, benefited from keeping in touch with their doctors and monitoring their medications. In addition, healthcare experts believe the approach is rather cost-effective. Some of these benefits included the elimination of the need to upgrade information infrastructure systems as clinical standards or reporting criteria changed, as well as more efficient data processing in relation to the growing amount of data and information collected from patients through electronic and personal health records. This is evident in terms of data storage and the number of servers required to handle such large amounts of data. Cloud computing is enabled by the use of mobile phones and tablets by medical professionals and patients to access healthcare facilities.

The evolution of Cloud Computing changing to the cloud has two advantages for the healthcare sector. Both healthcare professionals and patients have found it to be beneficial. On the business side, cloud storage has shown to be successful in lowering operational costs while enabling hospitals to provide high-quality, personalized care. Patients who have grown used to receiving treatment quickly are likely to get the same from the healthcare system. The following are some of the ways cloud consulting is affecting healthcare:

1. Cost-cutting is one of the first topics that springs to mind

    Cloud computing is based on the availability of computational services such as data storage and processing power on demand. The requirement for hospitals and healthcare providers to buy computers and servers outright is removed. There are no upfront costs involved with data collection in the cloud. You only pay for the services you use, resulting in significant cost savings.

    In addition, cloud computing offers the greatest ergonomic platform for scaling, which is a desirable advantage in today's world. With patient data streaming in not only from EMRs, but also from a range of healthcare applications and wearable's, a cloud-based environment proves to be suitable for scaling and undertaking capacitive redesign while keeping costs minimal.

2. Interoperability ease

    Regardless of where the data originate or are kept, interoperability aids in the creation of data linkages throughout the healthcare industry. As a consequence of interoperability driven by cloud usage, patient data are readily available for distribution and gathering knowledge to improve healthcare planning and execution.

    Cloud infrastructure helps healthcare providers to efficiently access medical data obtained from a range of sources, share it with key partners, and issue medications and medication protocols on schedule. It also decreases the gap between experts, allowing them to review cases and have opinions independent of geographical restrictions.

    Interoperability between the different sectors of the healthcare industry—pharmaceuticals, insurers, and payments—is also improved by keeping medical data in the cloud. This facilitates the smooth sharing of data between the multiple stakeholders, thus speeding up healthcare delivery and increasing quality.

3. Links to specialized analytical methods

Healthcare info, both organized and unstructured, is a tremendous resource. In the cloud, specific patient data from multiple sources can be collated and computed. The use of Big Data Analytics and AI algorithms on cloud-stored patient data may help advance medical research. Because of the cloud's improved computing capacity, processing large datasets becomes easier.

Analyzing patient data will also lead to the creation of more customized treatment strategies for patients on an individual basis. It also guarantees that every appropriate medical information is documented and that nothing is omitted when administering treatments. When it comes to extracting specific health data, cloud-based data mining comes in handy.

4. Data possession by the patient

Cloud computing decentralizes data and enables individuals to make health-related choices for themselves. It promotes patient interest in health-related choices and helps to better informed judgments by acting as a tool for patient understanding and participation.

When saving data in the cloud, patient data and medical pictures may be stored and retrieved promptly. Although cloud security is still an issue, cloud data storage is undeniably more reliable. With increased server uptime, data redundancy is reduced. Data recovery is much better because the backups are automatic because there isn't a single contact point where the data are stored.

5. Skills of telemedicine

The ability to view data from a distance is one of the most important benefits of cloud computing. The integration of cloud computing and healthcare has the ability to improve telemedicine, post-hospitalization treatment arrangements, and virtual drug adherence, among other healthcare-related features. Telehealth increases access to healthcare facilities as well.

Telemedicine applications make healthcare access more accessible while still improving the patient experience. Cloud-based telehealth services and software allow for simple data sharing, improved connectivity, and healthcare coverage for patients during the prevention, treatment, and rehabilitation phases.

- Information and Communication Technology (ICT) is a term that refers to the use of, in its most basic form, an electronic medium for producing, storing, controlling, receiving, and transmitting data from one location to another. It promotes message delivery by making it more comfortable, accessible, readable, and interpretable. Mobile phones, the Internet, cellular networks, computers, radios, televisions, satellites, base stations, and other devices are used. Knowledge is created, stored, communicated, transmitted, and managed using these tools.

6.8   INFORMATION AND COMMUNICATION TECHNOLOGY

ICT is a wide topic with many applications. It covers a wide range of topics concerning IT and how it impacts other areas of human effort. It is the most rapidly growing research sector and a sustainable source of income. It is the integration of telephone and computer networking in a single cabling device that makes for simple data collection, exploitation, control, and retrieval. Database administration, computer engineering, and software development are all subjects covered. Web design, mobile device creation, project management, security, networking research, media equipment, digital engineering, computer studies, the Internet, Internet protocol, intranet, application software, system software, signal technology, and base station management are only some of the jobs available.

ICT's Importance
- It strengthens students' critical brains, helping them to research and discover proper solutions to challenges in all relevant fields that use it as a learning aid.
- It encourages students to be creative and create innovative scientific answers to issues as a fresh experimental field of study.
- It enables the storing and retrieval of data.
- It improves computer networking, which is now referred to as the Internet and intranet.
- It boosts national economic development since it is a reliable source of revenue for all nations that recognize its significance (Figure 6.4).
- It creates lucrative employment and raises awareness of other issues, making it a viable source of income. ICT may be used in almost every area of teaching, such as in the classroom using a projector.
- It provides a forum for local and worldwide academics in the field of IT to exchange ideas and developments.
- It serves as a platform for e-learning as well as an online library. As a consequence, distributing information is easier than ever before,

FIGURE 6.4   Communication technology.

- That is essential to globalization in all of its forms, as well as the achievement of the United Nations' Millennium Development Goals established in 2000.
- It's used in various departments to keep records of government operations and administration.

6.9   TELEMEDICINE

Instead of personally attending a doctor's office or hospital, telemedicine helps patients to connect with a healthcare provider through technology. Using video, Web portals, and email, you can discuss conditions, medical problems, and more in real time from a healthcare professional. You can get a diagnosis, hear about the medical choices, and get a prescription through telemedicine. Healthcare providers can also remotely track readings from medical instruments and keep an eye on your condition if appropriate.

Telemedicine can be divided into three categories:

- Immersive medicine, also known as "live telemedicine," is a form of telemedicine in which doctors and patients collaborate in real time.
- Remote patient monitoring: This helps nurses to keep an eye on patients and use handheld medical devices to record data such as blood pressure and blood sugar levels.
- Store and forward: Providers may share a patient's confidential records with other physicians or experts.

Telemedicine has expanded to deliver treatment in a number of areas as technology has progressed. This includes personal physician-run websites, video tech that allows for remote appointments, and applications managed by telemedicine providers (Figure 6.5).

Telemedicine is not appropriate for emergency situations involving X-rays, splints, or casts, such as a heart attack or stroke, wounds or lacerations, or broken bones requiring X-rays, splints, or casts. Every condition that necessitates urgent, hands-on attention should be dealt with in person. Telemedicine, on the other hand, is suitable for minor problems and follow-up appointments.

For example, arrange a video meeting with your healthcare provider to address your symptoms if you think a cut is contaminated. If you're on break and believe you're having strep throat, you should call your primary care provider. If you want birth control, you can discuss the requirements and receive a prescription the next day.

It may be used for a number of other health needs, such as psychotherapy and tele dermatology, which includes consultations on moles, rashes, and other skin disorders. Other general conditions treated by telemedicine include colds and flu, bug bites, sore throats, diarrhea, and pink eye.

Telemedicine vs. telehealth: What's the difference

Although the words telemedicine and telehealth sound identical, they are not synonymous.

FIGURE 6.5 Telemedicine.

Telemedicine is described by the World Health Organization as "healing from a distance." This allows you to get medication without having to have an appointment with a doctor or visit their clinic for emergency care.

Encourage and promote the use of electronic information and networking technologies in long-distance clinical treatment, patient and professional health education, public health, and health administration. Telehealth isn't a commodity or a program. It's a method of bettering medical safety and specialist education. Telehealth encompasses nonclinical events such as appointment preparation, continued medical school, and specialist training, in addition to telemedicine.

Telemedicine has a number of advantages.

Telemedicine's rise is being fueled by the need for more affordable healthcare. Telemedicine can help people boost their physical health and well-being, whether they live in a rural location or have a busy life that keeps them from seeing a doctor.

As telemedicine becomes more common, many healthcare policies are starting to provide provision for telemedicine visits. In certain states, healthcare companies are expected to cover telemedicine visits at the same cost as in-person medical visits.

Telemedicine programs can be reimbursed by Medicaid if they fulfill the federal standards and availability criteria.

6.10 MOBILITY

Mobility is changing healthcare today. Smart devices and applications are not only empowering health agencies, but they are also greatly changing the patient/provider relationship. Providers can offer better treatment due to mobility. The use of mobility in healthcare is improving efficiencies, improving the patient experience both within and outside the hospital, and improving contact and coordination between patients and healthcare professionals, as well as between providers and their colleagues. Patient tracking using mobile devices is expected to save substantial healthcare costs in the immediate future.

Mobility and cloud access have significantly assisted in expanding connectivity for both patients and physicians. Instead of using paper charts, hospitals, insurance providers, and physicians' offices are also keeping patient care data in the cloud, with patients able to view examination results online 24 hours a day, 7 days a week.

Mobility has a range of advantages, including the treatment of post-discharge cases and the management of chronic illness cases. Similarly, for post-surgical treatment, patients can benefit greatly from mobility. This allows healthcare providers to reduce readmissions while still improving efficiency. The gathering of data from medical instruments on a regular basis assists doctors in obtaining useful knowledge and accurate information. This aids in dramatically reducing costs and improving patient quality.

Telemedicine has benefited greatly from the use of mobility and cloud connectivity in the healthcare ecosystem. Patients and healthcare services benefit from the freedom and convenience that telemedicine brings. Telemedicine has assured that a patient's treatment opportunities are not limited by their geographic position. They, too, will now enjoy the best level of treatment if they have access to the Internet and a smartphone. Telemedicine is cost-effective and time-saving. Patients no longer have to prepare their days around follow-up appointments. Instead, they will get the drug refill or check-up they need by dialing into a conference call.

Mobility has been a great enabler for Telemedicine, whether it's live streaming of medical data from a connected computer, high-fidelity audio communication over a handheld device, or information sharing from an Internet of Things (IoT) medical device. Telemedicine has shown to be more beneficial in the area of mental health than in any other medical field. Those in need of mental assistance will now connect with a psychiatrist at the touch of a button, for a fraction of the cost of a full office appointment.

In almost every area of the healthcare ecosystem, versatility can be a way of life for caregivers and patients. Because of the enormous simplicity of recording, exchanging, reviewing, and collaborating medical knowledge, the widespread use of mobility technologies is promoting faster implementation of telemedicine applications. An increasing variety of medical apps for mobile devices is assisting in the development of telemedicine for both doctors and patients. As a result, hospitals are becoming more effective, and patient conditions are improving.

6.11 ELECTRONIC HEALTH RECORDS

An EHR is a computerized representation of a patient's paper medical chart (EHR). EHRs are real-time, patient-centered records that make information readily and securely accessible to authorized users. While an EHR framework may give a patient's medical and prescription history, it is intended to go beyond the traditional health reports gathered in a provider's office to provide a more complete picture of a patient's care. EHRs are critical components of health IT because they may contain a patient's medical information, diagnoses, prescriptions, recuperation plans, vaccination dates, allergies, radiological pictures, and tests and test results, among other things. IoT provide doctors with access to evidence-based resources that they may utilize to make therapeutic choices: Provider process should be optimized and customized.

One of the most important features of an EHR is that it allows licensed physicians to create and preserve patient data in a digital format that can be shared with other professionals throughout healthcare companies. Because they include information about all practitioners involved in a patient's care, EHRs are intended to share information with all healthcare providers and organizations, including laboratories, doctors, diagnostic imaging facilities, pharmacy, emergency departments, and school and workplace clinics.

EHRs are patient-centered, real-time records that render details easily and securely available to approved users. While an EHR framework can provide a patient's medical and prescription history, it is designed to move beyond conventional health reports collected in a provider's office and to offer a more comprehensive picture of a patient's treatment.

EHRs can:

- archive a patient's medical history, diagnosis, prescriptions, recovery schedules, immunization dates, allergies, radiology images, and diagnostic and test results
- provide patients access to evidence-based resources for making clinical decisions
- simplify and streamline physician workflow

EHRs enable a bigger and more seamless interchange of data through a contemporary healthcare system that embraces and utilizes digital innovation and has the potential to alter how treatment is delivered and reimbursed.

Knowledge can be obtained anytime and wherever it is required with EHRs.

- Increased patient participation
- Improved care coordination
- Improved diagnostics and patient outcomes
- Practice efficiencies and cost savings

6.11.1 BENEFITS OF EHRs

EHRs have a range of advantages, including the potential to optimize patient treatment and simplify a multitude of activities for the practice. They also urge physicians to exchange files remotely and in real time with one another, ensuring that every clinician dealing with a patient has access to the most up-to-date, complete, and trustworthy file possible.

- Better quality of care

    EHR enables doctors to provide excellent service to their patients by making patient information easily accessible, resulting in more effective treatment. They also improve treatment effectiveness while increasing clinical productivity in the practice. Most EHRs include health analytics that assist physicians in identifying patterns, forecasting diagnoses, and recommending treatment choices. Instead of depending on trial-and-error methods, these analytics provide more efficient overall patient outcomes the first time around. Patients may visit patient websites to view previous medical data such as lab and imaging results, medications, diagnoses, and other information. Patients may communicate with their physicians by sharing information, sending text messages, or even participating through video conferencing.
- Willingness to communicate with one another

    Combining EHR with other systems such as EMR may help hospitals improve treatment quality. When patients continue to visit physicians, manage chronic illnesses such as diabetes, or plan to move to a home health or hospice environment, an interoperable EHR system is critical.
- Increased productivity

    EHR, as previously said, allows doctors to offer more precise care and diagnosis while still saving time. They reduce wait times for appointments and office visits by maintaining a patient-centered approach, allowing health providers to treat more patients on a regular basis.

6.11.2 USE OF MOBILE AND WIRELESS TECHNOLOGY TO IMPROVE EHR INDUSTRY

A corporation despite being reluctant to engage in IT, the healthcare sector was a pioneer in the introduction of smartphone and wireless technologies, likely due to the constant urgency of interactions between physicians and nurses who are almost constantly on the move. Doctors, for example, were among the first large groups to implement personal digital assistants in the office. The following are several other popular applications of wireless technologies in the healthcare field:

- Offering a way for patients to view and monitor their EHRs at their bedsides to ensure that patient data are reliable and up to date.
- Allowing nurses to check barcodes on patient wristbands and prescriptions to assist them in administering the correct prescription in the required dose at the appropriate time (A wireless network links an attached device on a neighboring cart to a directory of physician prescription orders.)
- Using mobile devices to connect with healthcare workers no matter where they are.

Electronic IT and networking are used by telehealth workers to facilitate remote health services, offer clinical and patient health-related training, and help with healthcare management. Internet, cable and cellular technologies, desktop and tablet computers, videoconferencing, digital media, and store-and-forward, high-resolution imagery are all common telehealth technologies. Thousands of mobile phone applications are available to increase patient access to healthcare details and enable physicians to track their patients' symptoms. Text messaging-based applications include appointment and prescription reminders, medication and vital sign monitoring, and food and weight management. One iPhone app, for example, will take your blood pressure and heart rate, timestamp and register the details, and then send it to a doctor. However, there are several possible risks to using smartphone devices. Medical professionals do not interact with patients through SMS text messages, according to the Joint Commission on Accreditation of Healthcare Organizations. If confidential patient information is transmitted through regular text messaging, HIPAA regulations can be compromised. This ruling is pushing mobile app developers to drop SMS text messaging in favor of a more reliable contact form. Telemedicine is a form of telehealth that helps patients to access medical services from healthcare providers that are not in the same place as them. This technology helps healthcare providers to accommodate more patients in a wider geographic region, reducing the need for patients to travel for treatment.

Telemedicine is divided into three types: store-and-forward, live telemedicine, and remote surveillance. Obtaining data, voice, pictures, and video from a patient and then forwarding it to a medical expert for later assessment is known as store-and-forward telemedicine. This method of supervision would not necessitate the patient and care provider being present at the same time. Getting access to such data, on the other hand, will enable healthcare practitioners to spot complications and interfere with remote patients before life-threatening circumstances arise.

6.12   TELEHEALTH

Telehealth is a term that refers to the use of technology to

Patients and caregivers are divided by distance in the delivery of healthcare services. Telehealth makes use of information technology to diagnose and treat diseases and impairments, as well as perform tests and assessments and offer continuing education to healthcare providers. Telehealth can continue to achieve universal health coverage by improving patients' access to high-quality, cost-effective healthcare no matter where they are. It is particularly helpful to those who live in rural areas, disadvantaged populations, and the elder persons.

Telemedicine and Telehealth, in layman's words, refer to the transmission and sharing of medical data between various locations. Telemedicine, as well as telehealth, are characterized by the sharing of still photographs. According to the American Telemedicine Association, patient consultations through video conferencing, patient websites, remote monitoring and continuing medical education, monitoring of vital signs, patient-focused wearable software, nursing call centers, and several more applications are all possible.

Telehealth has been around since the 1960s, with one of the first uses being the tracking of astronauts' physiological parameters. Several technical and communications resources have been introduced across the years, due to technological advances, to allow the transfer of patient information for referrals and appointments in nearly any medical community and specialization. Telehealth systems for patients have also allowed for remote care management, customer health contact and statistics, and provide medical education. Some of the most common distribution methods include networked programs that connect tertiary medical centers to outlying centers and clinics in remote communities, home phone–video connections, point-to-point connections to hospitals and clinics, Web-based e-health service sites, and home tracking links.

For a while, however, the risk of embracing and participating in telehealth systems was too great, and the delivery of telehealth resolutions and hospital-based networks was too expensive. However, as a result of technology advances, enhanced broadband networks are both efficient and inexpensive, resulting in a higher return on investment in telehealth than ever before. Telehealth systems can be used to link clinicians with patients in different locations through real-time audio and video in almost all medical specialties. In other instances, treatment centers may gather data remotely and send it to a central tracking system for analysis using telemedicine.

Outlying healthcare services are often required to move patients prematurely or to refer patients to difficult cases that are outside the knowledge base of local providers. As a result, patients are often moved through long distances for direct treatment or telehealth consultations with specialists. These referrals and transfers can be expensive for the patient because they pose health, technological, and financial problems for all parties involved. Telemedicine has the potential to address these problems, minimize travel frequency, and have significant efficiencies and returns for all stakeholders concerned.

6.13   MEDICAL INFORMATION WEBSITES FOR LAY PEOPLE

Good individuals, as well as those who are sick, need detailed knowledge about a wide variety of medical issues in order to achieve a deeper understanding of healthcare systems and take full control of their health. Obviously, laypeople cannot become as well-informed as qualified medical professionals, but the Internet provides a wealth of healthcare knowledge. These websites have a vital duty to provide authentic, up-to-date, and impartial information. Html, illustrations, and photographs on a medical knowledge website are included solely for educational purposes. These websites are not designed to replace legitimate medical care, diagnosis, or care. For any concerns about a medical condition, people should still follow the opinion of a physician or other competent healthcare professional.

Because of what he or she reads on medical records website, a patient should never ignore legitimate medical advice or avoid finding it. Some hospital providers, employers, and medical insurers provide valuable online resources that reach beyond basic health records in addition to freely accessible information on the Internet.

Patients may use these measures to compare the efficiency, protection, and cost of hospitals around the world. There are also risk factors for individual healthcare options, as well as national average prescription and treatment costs. Many of these websites also offer information about the coverage and prices of services provided by in-network and out-of-network healthcare providers.

A person who wants a hip replacement, for example, can go online to learn more about the procedure, as well as possible care choices, a list of questions to ask the doctor, possible risks, local hospitals that administer the procedure, and hospital quality-of-service statistics, such as the amount of recorded postoperative infections and other complications.

REVIEW QUESTIONS

1. List the advantages and disadvantages of Teleworking for employees.
2. Discuss the most important legal questions around the usage of social networking websites.
3. What is E-Rate program and what are the problems with it?
4. Write short notes on worker's productivity, cyberbullying, and cyberstalking, and crime in virtual world.
5. Discuss the impact of IT on Healthcare Costs.
6. "There is no impact of IT on quality of life." Do you agree with the statement? Justify your answer.
7. Explain Electronic Health Record and Telemedicine.
8. Discuss the impact of Information Technology on standard of living and worker productivity.

MULTIPLE CHOICE QUESTIONS (MCQs)

1. **What is an example of a more sophisticated approach for entering data into an EHR?**
   a. Use a voice recognition system
   b. Typing with a keyboard
   c. Use an iPod
   d. Use of a transcriptionist
2. **Which one of the following characteristics of EHRs is not a good attribute?**
   a. Increased quality of patient care
   b. Decreased efficiency and savings
   c. Access to health records is becoming easier to come by
   d. Improved patient safety
3. **Which of the following is a stumbling block to EHR implementation?**
   a. Lack of standards
   b. E-prescribing options
   c. Identified ROI
   d. Increased patient safety

4. **A protection function that controls information access depending on the type of data each user wants to see or alter.**
   a. dashboard
   b. access level
   c. lookup
   d. E/M codes

5. **A device developer is someone who ensures that applications are installed on schedule, on budget, and with reasonable consistency.**
   a. project manager
   b. systems designer
   c. external system user
   c. systems owner

6. **For an information system, which of the following is not a business driver?**
   a. Security and privacy
   b. Proliferation of networks and the Internet
   c. Control of information properties
   d. Redesigning corporate processes

7. **Which of the following is not an information system's technological driver?**
   a. Object technologies
   b. Enterprise applications
   c. Collaborative technologies
   d. Knowledge asset management

8. **The typical information transformation cycle is as follows:**
   a. Data to information to knowledge
   b. Information to data to knowledge
   c. Knowledge to data to information
   d. Data to knowledge to information

9. **Monitoring a company's compliance restrictions necessitates an examination of**
   a. a company's customers
   b. a company's macro-environment
   c. a company's outputs
   d. a company's micro-environment

10. **Information systems that assist business activities in interacting with suppliers are known as**
    a. decision support systems
    b. expert information systems
    c. front office information systems
    d. back office information systems

ANSWERS TO MCQs

Q1: a, Q2: b, Q3: b, Q4: d, Q5: a, Q6: b, Q7: a, Q8: a, Q9: c, Q10: d

7 Case Studies

LEARNING OBJECTIVES

- Recommend controls to avoid an enterprise security breach.
- To create awareness among the users for any Phishing or Smishing cases
- Recommend controls to avoid an enterprise data breach.

INTRODUCTION

Most regions of the globe are surrounded by a high spread of information technology. This opens up opportunities for IT companies operating both locally and worldwide. Sensitive data is also sent through the internet during transactions and conversation. The technology revolution has transformed the globe, transforming computers and the internet into fundamental communication devices that have transformed the world into a village. They also utilize computers to store data and analyze both private and public data. Cyber society is a virtual world created by technology, and it is the job of information technology and telecommunication engineering to propel the world forward. However, extra effort will be needed to establish strong legislation in order to prevent cybercrime.

Another concern is cyber terrorism, in which terrorists are enticed to use the internet and computer network systems to carry out their operations. They communicate over the internet, which makes money laundering easier. Because the majority of cases stem from computers and related technology, such as hackers, stolen sensitive information, and the use of electronic commerce to promote business and communications, cyber law will be required. This implies that information transmitted electronically is extremely sensitive, necessitating the use of cyber law to provide and disapprove of facts in question.

To create awareness among such types of attacks such as data breaches, security breaches, money extortion, certain actual incidents happened in some organizations as well as cyber police station, Nagpur City, MS, India, are identified as a part of case studies to prevent the users for being the victim for any such kind of attacks. The preventive steps to avoid such types of attacks at the end of each case study are highlighted in this chapter.

Case Study 7.1

SECURITY BREACH

A security breach is any incident that results in unauthorized access to computer data, applications, networks or devices. It results in information being accessed without authorization. Typically, it occurs when an intruder is able to bypass security mechanisms.

DOI: 10.1201/9781003280989-7

Technically, there's a distinction between a security breach and a data breach. A security breach is effectively a break-in, whereas a data breach is defined as a cybercriminal getting away with information. Imagine a burglar; the security breach is when he climbs through the window, and the data breach is when he grabs your pocketbook or laptop and takes it away.

Confidential information has immense value. It's often sold on the dark web; for example, names and credit card numbers can be bought and then used for the purposes of identity theft or fraud. It's not surprising that security breaches can cost companies a huge amount of money. It's also important to distinguish the security breach definition from the definition of a security incident. An incident might involve a malware infection, DDOS attack or an employee leaving a laptop in a taxi, but if they don't result in access to the network or loss of data, they would not count as a security breach.

7.1.1 Examples of a Security Breach

When a major organization has a security breach, it always hits the headlines. Security breach examples include the following:

- In the case of Equifax in 2017, a website application vulnerability caused the company to lose the personal details of 145 million Americans. This included their names, SSNs, and drivers' license numbers. The attacks were made over a three-month period from May to July, but the security breach wasn't announced until September.
- Yahoo-three billion user accounts were compromised in 2013 after a phishing attempt gave hackers access to the network.
- eBay saw a major breach in 2014. Though PayPal users' credit card information was not at risk, many customers' passwords were compromised. The company acted quickly to email its users and ask them to change their passwords in order to remain secure.
- Dating site Ashley Madison, which marketed itself to married people wishing to have affairs, was hacked in 2015. The hackers went on to leak a huge number of customer details through the internet. Extortionists began to target customers whose names were leaked; unconfirmed reports have linked a number of suicides to exposure by the data breach.
- Facebook saw internal software flaws lead to the loss of 29 million users' personal data in 2018. This was a particularly embarrassing security breach since the compromised accounts included that of company CEO.
- Marriott Hotels announced a security and data breach affecting up to 500 million customers' records in 2018. However, its guest reservations system had been hacked in 2016 the breach wasn't discovered until two years later.
- Perhaps most embarrassing of all, being a cyber security firm doesn't make you immun-Czech company Avast disclosed a security breach in 2019 when a hacker managed to compromise an employee's VPN credentials. This breach didn't threaten customer details but was instead aimed at inserting malware into Avast's products.

A decade or so ago, many companies tried to keep news of security breaches secret in order not to destroy consumer confidence. However, this is becoming increasingly rare. In the EU, the GDPR (General Data Protection Regulations) require companies to notify the relevant authorities of a breach and any individuals whose personal data might be at risk.

7.1.2 TYPES OF SECURITY BREACHES

There are a number of types of security breaches depending on how access has been gained to the system:

- An **exploit** attacks system vulnerability, such as an out-of-date operating system. Legacy systems which haven't been updated, for instance, in businesses where outdated and versions of Microsoft Windows that are no longer supported are being used, are particularly vulnerable to exploits.
- **Weak passwords** can be cracked or guessed. Even now, some people are still using the password 'password', and 'pa$$word' is not much more secure.
- **Malware attacks,** such as phishing emails can be used to gain entry. It only takes one employee to click on a link in a phishing email to allow malicious software to start spreading throughout the network.
- **Drive-by downloads** use viruses or malware delivered through a compromised or spoofed website.
- **Social engineering** can also be used to gain access. For instance, an intruder phones an employee claiming to be from the company's IT helpdesk and asks for the password in order to 'fix' the computer.

7.1.3 SECURITY BREACH INVESTIGATION AT IT NETWORKZ, NAGPUR

IT-NetworkZ is a technology firm that works in the field of information technology. IT-NetworkZ began its operations by providing information technology services such as IT Infrastructure Management, Professional IT training, and online assessment facilities. Company's headquarter is located in Nagpur, Maharashtra State in India. There was a team of 17–18 employees, and one of the Company's 23-year-old employees attempted for breaking security all the terminals. He has installed Keylogger software in all the terminals of the organization connected with single server computer. The credentials of the users of company have been diverted to this employee's through the Keylogger. He placed key logger software on all of the company's computers, so that anytime any employee, even the CEO, types something on the keyboard, it is immediately stored on the server computer. As this employee has access to the server computer, all data with user name and passwords were in hidden file (Figure 7.1).

He was very much interested to know information in personal mails and official communications and preferred to trap all user names and passwords; he was misusing other employees' credentials, such as personal and official email addresses, social media accounts, and so on. Meanwhile, several employees who believed their

LETS GET INTERNETWORKED

FIGURE 7.1   Logo of IT NetworkZ, Nagpur.

personal accounts were being utilized by someone; employees filed a complaint with the company's operations manager, Mr. Deepak Dhote.

After a thorough investigation, Mr. Deepak Dhote and his security team discovered the perpetrator within his business. Putting restrictions on network access and by introducing additional security applications and hardware firewalls in system and organization tried to avoid a similar incidence. Following the event, the Operations Manager advised all workers to keep all facts private until further notice. This employee was discovered to be the perpetrator for violating the company's security and this candidate was a part of the IMS team, worked for more than 1.5 years and had easy access to all systems and server. After official investigation, everything was open and the candidate realized that there was no space to escape. One thing which was unknown to all was the reason why he did this, which he kept changing (Figure 7.2).

This security breach incidence was happened due to:

- Unauthorized information access, use, disclosure, alteration, or destruction.
- Interference with the operation of information technology laws.
- Intentional or unintentional violation of the acceptable usage company's policy.

The following are some examples of IT security incidents in the various IT organizations:

- Hacking into a computer system
- Unauthorized use or access to systems, software, or data
- Changes to systems, software, or data that are not permitted
- Equipment used to store or work with sensitive university data is lost or stolen.
- A denial-of-service (DoS) assault

FIGURE 7.2   Investigation of security breach attack at IT NetworkZ, Nagpur.

- Disruption of the intended usage of IT resources
- User accounts that have been hacked

7.1.4   PREVENTIVE MEASURES TO BE TAKEN TO AVOID SECURITY BREACH INCIDENCES

In such cases, providing detailed information with respect to the following points is essential.

(a) Name of sufferer (b) Department (c) Electronic mail address (d) Contact information (e) An explanation of the IT security issue (f) The date and time when the issue was first discovered (if possible) (g) The affected machine's IP address (h) Any other resources that may be impacted like synchronized mobile phones, tab, etc.

The concerned unit of the organizations' security team will contact the sufferer, and this team will design a plan for further containment and mitigation.

7.1.5   CONCLUSIONS

Mr. Deepak Dhote, Operations Manager of IT Networkz, Nagpur offered some tips on how to avoid such type of attacks:

- Maintain your composure. There is a system in place for dealing with issues.
- Don't put speed ahead of accuracy. Don't make hasty decisions.
- Involve your leadership as soon as possible. Remind them that all information should be confined to a need-to-know basis, especially early in the inquiry.
- Each and every detail is crucial. Inform the ITS investigation coordinator immediately.

Case Study 7.2

DATA BREACH

To define data breach: a data breach exposes confidential, sensitive, or protected information to an unauthorized person. The files in a data breach are viewed and/or shared without permission.

Anyone can be at risk of a data breach – from individuals to high-level enterprises and governments. More importantly, anyone can put others at risk if they are not protected.

In general, data breaches happen due to weaknesses in:

- Technology
- User behavior

As our computers and mobile devices get more connective features, there are more places for data to slip through. New technologies are being created faster than we can protect them. Devices in the IoT sector are proof that we are increasingly valuing convenience over security. Many "smart home" products have gaping flaws, like lack of encryption, and hackers are taking advantage.

Since new digital products, services, and tools are being used with minimal security testing, we shall continue to see this problem grow in the near future.

However, even if the backend technology was set up perfectly, some users are likely to still have poor digital habits. All it takes is one person to compromise a website or network. Without comprehensive security at both the user and enterprise levels, devices and the information are almost guaranteed to be at risk.

7.2.1 PROTECTING THE ORGANIZATION FROM DATA BREACH OCCURRENCES

- An Accidental Insider. An example would be an employee using a co-worker's computer and reading files without having the proper authorization permissions. The access is unintentional, and no information is shared. However, because it was viewed by an unauthorized person, the data is considered breached.
- A Malicious Insider. This person purposely accesses and/or shares data with the intent of causing harm to an individual or company. The malicious insider may have legitimate authorization to use the data, but the intent is to use the information in nefarious ways.
- Lost or Stolen Devices. An unencrypted and unlocked laptop or external hard drive- anything that contains sensitive information that goes missing.
- Malicious Outside Criminals. These are hackers who use various attack vectors to gather information from a network or an individual.

This case study is also from the same organization i.e. IT-NetworkZ, Nagpur, which is a technology firm that works in the field of information technology.

One of the company's 44 year old employees was identified as the perpetrator of the data breach. This employee was on a Technical Head position and had easy access to vital information and he was aware about all most all business deals and activities. Exact and unique data was identified on competitor's application which brought management on toe and an inquiry started.

A data breach occurs when sensitive, confidential, or otherwise protected data is accessed and/or disclosed without authorization. Data breaches may happen in any size of business, from tiny startups to large multinationals. Personal health information (PHI), personally identifiable information (PII), trade secrets, and other private information may be included.

Personal information, such as credit card numbers, Social Security numbers, driver's license numbers, and healthcare records, as well as business information, customer lists, and source code, are all common data breach targets.

A data breach occurs when someone is not allowed to see or take personal data from the entity in responsibility of securing it. If a data breach results in identity theft and/or a violation of government or industry compliance regulations, the guilty company may suffer penalties, lawsuits, reputational damage, and even the loss of its business license. For monetary benefit, a working employee of the company has given out all of the company's information to other rivals. As a business employee, he has complete access to the corporation.

He resigned from the firm immediately after misusing the organization's private data as his deed was an open secret to all staff members and management. He worked more than two years and this unexpected incident occurred, management somehow managed to get everything from him and succeed to control the loss.

In both the cases, Non Disclosure Agreement (NDA) was taken from the employee but in such cases greediness of a human being may wins over rules/regulations.

7.2.2 CONCLUSIONS

Mr. Deepak Dhote, IT NetworkZ organization's Operations Manager, suggested the following preventive steps to avoid data breaches of this nature:

There is no single security instrument or control that will completely eliminate data breaches. Commonsense security procedures are the most reasonable method of preventing data leaks. These include well-known security fundamentals like:

- carrying out regular vulnerability evaluations
- testing for intrusion
- putting in place malware security that has been proved to work
- Using secure passwords/passwords
- Deploying software fixes to all systems on a regular basis

Case Study 7.3

EXTORTION OF MONEY

Extortion is the wrongful use of actual or threatened force, violence, or intimidation to gain money or property from an individual or entity. Extortion generally involves a threat being made to the victim's person or property, or to their family or friends.

The fraud takes place after a fake account is created using the profile picture of a real user. The fraudsters will then contact the friends of the real user and ask them for money. Several such cases have been reported.

Usually, the money is asked on the pretext of an emergency. Those who send the money online will realize that they have been cheated only when they contact the real person.

7.3.1   Incidence Reported by Cyber Police Station, Nagpur City, MS

This money extortion case was registered at GITTIKHDAN Police Station in Nagpur city of Maharashtra State in India. The was reported as CRIME NO 554/2020 SEC 419, 420, 468,471,120, B, 34 IPCR/W 66 (C) IT ACT, under SEC 12 (C) PASSPORT ACT 1967. This case has been investigated by Mr. Vishal Mane and his team under the guidance of Dr. Ashok Bagul, Senior Police Inspector at Cyber Police Station, Nagpur, Maharashtra, India.

In the said case, modus operandi of the accused is regarding the extortion of money wherein the accused lured the complainant in the name of gifts and parcels and also promised for job in return of money whereby extorted the huge amount of Rs. 4,240,069. The complainant in the present case is Mr. Ivol Singh Samser Singh having around 56 years of age. In the instant matter, extortion and cheating of money were carried out by the attacker. The platform of Facebook and WhatsApp was used for the said cheating.

Initially, the cyber attacker made friendship with the complainant's wife thereby establishing the confidence and trust of the victim. Thereafter he told her that one consignment of gift and parcel in exchange of money gave various reasons of custom charges, transfer charges and many more fake and fraudulent ways for extortion of money to the wife of complainant's. Also, one lady officer posing herself as custom officer, New Delhi called the victim for clearing the said consignment, they have to make payment otherwise an offense will be registered against her. Due to fear of the same, complainants wife out of trust transferred the total amount of Rs. 4,240,069 between during 2020. The said amount has been transferred to various bank accounts of the attacker. The bank details and transfer receipt used were shared on Facebook and WhatsApp and a lot of data shared with the attacker has been recovered during the investigation.

The FIR in the said context was registered in 2020. Thereafter total of five accused were arrested from New Delhi. During the course of investigation, the cyber police team arrested four foreigners and one local person.

7.3.2   Conclusions

Nowadays similar types of money extortions are being reported regularly. Dr. Bagul, Sr. PI of cyber police station suggested the following measures to be taken to avoid money extortions.

A. Don't open any email or attachment from strangers.
B. Regularly monitor your bank account and credit report for any suspicious activity.

C. Use strong passwords and avoid using the same password for multiple websites.
D. Never give out any personal information via email or social media.
E. Adjust your social media security settings to provide the highest level of protection.
F. Before entering any personally identifiable information on a website, please ensure that the site is using "https" or the status bar which displays a "lock" icon.

Index